Masonry Wall Construction

Masonry Wall Construction

A. W. Hendry and F. M. Khalaf

London and New York

First published 2001
by Spon Press
11 New Fetter Lane, London EC4P 4EE

Simultaneously published in the USA and Canada
by Spon Press
29 West 35th Street, New York, NY 10001

Spon Press is an imprint of the Taylor & Francis Group

Typeset in 10/12 Sabon 3B2.603d [Advent] by
Keyword Publishing Services Ltd.
Printed and bound in Great Britain by
TJ International Ltd., Padstow, Cornwall

British Library Cataloguing in Publication Data
A catalogue record for this book is available
from the British Library

Library of Congress Cataloging in Publication Data
Hendry, A.W. (Arnold W.), 1921–
    Masonry wall construction / A.W. Hendry and F.M. Khalaf.
        p. cm.
    Includes bibliographical references and index.
        1. Brick walls–Design and construction.   2. Stone walls–Design and construction.   3.
    Masonry.   1. Khalaf, F.M. (Fouad M.), 1950–   II. Title.

    TH2243 H46 2000   00-039499
    693.1–dc21

ISBN 0-415-23282-1 (pbk.: alk. paper)

Contents

Acknowledgements

Thanks are due to the following for permission to reproduce illustrations used in this book, as identified in the text:

CERAM Building Technology. http://www.ceram.co.uk E-mail: info@ceram.co.uk
The Brick Development Association. E-mail: brick@brick.org.uk
KLB Klimatleichtblock GmBH. http://www.klb.de
JUWO Poroton-Werke. http://www.juwoe.de E-mail: vertrieb@juwoe.de

Extracts from BS 5628: Part 1 1992 and BS 1200 Amd 4510 1984 are reproduced under licence number 2000SK/0131. Complete editions of these standards can be obtained by post from BSI Customer Services, 389 Chiswick High Road, London W4 4AL.

Preface

Masonry in the form of brickwork, blockwork and natural stone is one of the most familiar construction materials. Perhaps it is for this reason that it tends to be taken for granted and that there is a feeling that little thought needs to be given to the design and building of masonry walls. This of course is far from the case as such walls have to perform simultaneously many functions including structure, thermal and sound insulation and weather protection as well as division of space. Each function has its own criteria all of which have to be met. Additionally, durability, cost and construction factors have to be considered so that masonry wall design and construction is in fact quite a complex matter.

Many books on the subject deal in depth with a particular aspect such as structure or materials. The present book on the other hand sets out to give a comprehensive view of masonry wall construction for students preparing to enter one or other of the professions related to building. It will also be relevant to the needs of recent graduates preparing for professional examinations and to others who require a general knowledge of this form of construction.

The aim has been to treat each aspect in sufficient detail to enable the reader to appreciate the problems involved and their treatment in practice. The text, starting with a historical perspective, outlines current methods in design and construction whilst seeking to identify future trends.

Considerable reliance has been placed on information derived from Building Research Digests and other publications of the Building Research Establishment as well as those from the Brick Development Association and other industry based organisations. These are listed after each chapter as items for further reading.

Arnold W. Hendry
Fouad M. Khalaf

Chapter 1

Introduction

1.1 Historical perspective

The archaeological and historical record over several millenia shows that the basic construction materials used to create human shelter were the same in all parts of the world and derived from the earth or from plant life, used as locally available and depending on climatic conditions. Figure 1.1 shows an example of a Stone Age house at Skara Brae in the Orkney Islands (*c*.2500 BC).

As societies became more complex, so the requirement for buildings and public facilities progressed beyond the need for rudimentary shelter. This gave rise to the evolution of masonry construction from rather crude assemblages of small stones or stone slabs jointed with mud to very large structures built from carefully shaped blocks. Thus by 2700 BC the Egyptian King Zoser had constructed the famous stepped pyramid at Sakkara. Many other pyramids were built by the Pharaohs in the following centuries, as tombs for themselves, without the use of cranes, pulleys or lifting tackle but of course with immense expenditure in labour and material. The Egyptians also built numerous temples of much greater complexity than the pyramids, demonstrating a very high degree of engineering skill in masonry construction.

Other societies in the Middle East used masonry and in particular the Sumerians developed the manufacture of clay bricks using moulds as early as 3000 BC. These bricks were generally sun dried although fired clay was known and used for the outside of important buildings. A later example of early brick construction on a large scale is King Nebuchadnezzar's Palace in Babylon on the River Euphrates in Iraq, built about 600 BC (Fig. 1.2).

The Greek civilisation used essentially the same techniques as the Egyptians, using stone walls, piers and lintels, but built much more sophisticated palaces and temples. The Romans, their successors as the dominant civilisation in the Middle East and Western Europe, were even greater builders and there are examples of their buildings and structures for almost every purpose. They used stone and fired clay bricks and a most significant development was their invention of pozzolanic cement made by the addition of sandy volcanic ash to lime mortar. This was used both for concrete and for mortar, yielding material of great strength and extending the application of masonry construction to aqueducts, bridges and domes as well as walled buildings exemplified by the basilica at Trier, Germany (Fig. 1.3) built in AD 300.

Roman building continued in the Byzantine or Eastern Empire especially with the construction of churches, some on a very large scale such as Hagia Sophia in

Figure 1.1 Skara Brae in Orkney. A domestic building interior in a New Stone Age settlement of the third millenium BC.

Figure 1.2 Side view of King Nebuchadnezzar's Palace in Babylon, Iraq, built about 600 BC. Restored at the end of the 20th century.

Constantinople (now Istanbul) in the 6th century AD. Through the Middle Ages the most notable masonry buildings in Europe were churches, cathedrals and castles. The latter ranged from vast complexes with formidable defensive walls to relatively small towers in disputed areas. An example of a border tower is Borthwick Castle, near Edinburgh (Fig. 1.4), built in 1430 with walls over 4 m thick. Such massive construction was necessary for their purpose but showed no great technical finesse and eventually they were unable to withstand the battering from ever more powerful cannons, and after their military value came to an end, some were converted into residences. An example of such a conversion is Dalhousie Castle (Fig. 1.5), also near Edinburgh,

Figure I.3 The Roman basilica at Trier, Germany, Built in AD 300 by Constantine the Great. Restored to its original form in the 19th and 20th centuries and now used as a church.

Figure I.4 Borthwick Castle, near Edinburgh, is an example of a late medieval tower house built in 1453 to afford protection from marauders. Its structure is largely unaltered and it has been in continuous occupation for over 500 years and is now a hotel.

which was built in the 12th century and converted into a country house in the 19th century and is still in use as a hotel. More often, however, the landowner built a new mansion or palace on or close to the site of the original castle, leading to a new style of masonry building, sometimes on a very grand scale. Many 'stately homes' of this kind

Figure 1.5 Dalhousie Castle, Midlothian, originally built in the 12th century, is even older than Borthwick Castle but has been completely altered and in the 19th century reconstructed as a country house. Now used as a hotel.

were built between the 16th and 19th centuries and are now valued as part of the national heritage.

Cathedrals built between the 11th and 14th centuries reached an extremely high technical and aesthetic level of building using stone masonry, as may be seen from Fig. 1.6 which shows the interior of Durham Cathedral.

In the 18th and 19th centuries the Industrial Revolution took place in Europe, the population expanded rapidly and towns and cities grew proportionately. This resulted in building on a previously unprecedented scale. With the exception of some industrial building, this was almost entirely in masonry and timber until near the end of the 19th century. Some of the building in towns and cities was of a high standard but housing for the new working class was generally not and rapidly deteriorated into slums.

A great deal of construction was required for new industrial development and transport infrastructure through this period, again built largely in masonry but increasingly with cast and wrought iron and eventually in steel and concrete. The latter materials have in the 20th century displaced masonry for many purposes but masonry has retained a predominant position in relation to low and medium rise housing and for non-structural purposes in buildings in which the structure is steel or concrete.

The earliest high rise buildings were in fact built in masonry. One of the early 'skyscrapers' in Chicago, was the Monadnock Building (Fig. 1.7), now preserved for its historic interest. This building is 16 storeys in height and of brickwork.

Figure 1.6 A view of the interior of Durham Cathedral, begun in the year 1093 and thus in use for almost 1000 years.

Figure 1.7 The Monadnock Building, Chicago, built in 1896. Represents the ultimate in high rise brickwork designed according to conventional rules, resulting in walls 1.8 m thick at ground level.

Figure I.8 Borough High Street Flats, London. Mid 20th century planning ideas in the UK led to the construction of high rise residential buildings in brickwork. Research made it possible for these to be built to 18 storeys with bearing walls no thicker than 225 mm.

One of the main factors permitting the adoption of such buildings was the development of reliable lifts but, as the wall thickness at the ground floor level of the Monadnock Building was 1.8 m, it was clear that, at least with the structural rules governing wall thickness available in the 1890s, this was not a practical proposition because of the loss of useable space at the lower levels. Thus little more was heard of high rise masonry buildings until the 1950s when apartment buildings in brickwork of up to 18 storeys were built, first in Switzerland, with a structural wall thickness of 150 mm or less. This was possible on the basis of research work at the Swiss Federal Institute of Technology. Similar work undertaken in the UK at the Building Research Station led to the publication of a code of practice for bearing walls in 1948 and a decade or so later to the construction of high-rise buildings such as the apartment building shown in Fig. 1.8.

Extensive research in many countries has led to the development of more sophisticated codes of practice but in recent years masonry has been used almost entirely for low to medium rise buildings and for cladding of steel and concrete buildings and the potential for high rise construction has not been fully explored. Figures 1.9–1.12 show examples of late 20th century masonry buildings and the use of the material as cladding.

Figure 1.9 Clinton Grange apartment block, Edinburgh. Cavity walls with concrete block inner leaf and facing brick outer leaf.

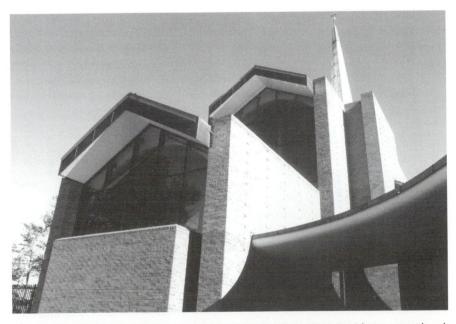

Figure 1.10 St. Barnabas Church, Dulwich, London. An interesting contrast with masonry church construction of earlier centuries. (Courtesy The Brick Development Association and H.O.K. Architects.)

Figure 1.11 Offices in Sunderland. A brick building constructed with a view to compliance with the requirements of sustainability. (Courtesy The Brick Development Association and Marshall Clay Products Ltd.)

Figure 1.12 Two interpretations of natural stone cladding. (a) Crowne Plaza Hotel, Edinburgh. The structure is large panel concrete with stone masonry 150 mm thick as the outer leaf of a cavity wall. (b) Scottish National Museum, Edinburgh. Sandstone slabs, 75 mm thick, individually attached to the concrete structural wall.

1.2   Applications of masonry wall construction

As discussed in the preceding section, until about one hundred years ago masonry was the dominant construction material for buildings of any size. In the 20th century, however, it has been displaced for many applications by steel and concrete but, remains of great importance for low and medium rise buildings, for cladding of buildings and for internal, non-loadbearing walls where the structural function is met by one or other of these newer materials.

The market for masonry construction may be divided into housing and non-housing sectors, the latter including industrial, commercial and educational buildings as well as a wide variety of buildings used for administrative and recreational purposes. In addition, there is a limited use of masonry construction for infrastructure, for example for retaining walls. In all sectors there is a significant requirement for masonry in the repair and maintenance of existing buildings.

Masonry walls may be external or internal and may be loadbearing, providing the structure, or non-loadbearing, sub-dividing space or acting as the cladding of the building. The design criteria to be satisfied in these applications differ although in many cases the same wall is required to meet more than one function; thus a loadbearing wall may also have to enclose space and provide sound insulation and fire protection. Similarly, in low-rise construction, as in housing, the outer walls have such combined functions and in both cases external walls must provide weather exclusion and define the appearance of the building.

Reinforced and post-tensioned masonry overcome the low tensile strength of the material and extend its use in situations where considerable lateral forces have to be resisted. The most obvious application of these techniques is to buildings located in seismic areas but they are also relevant where non-loadbearing panels are subject to substantial wind loads. Walls of cellular or T cross-section are particularly suitable for large, single cell buildings and in this case the potential for the adoption of such walls is greatly extended by post-tensioning.

1.3   Advantages of masonry construction

The first advantage of masonry wall construction lies in the fact that a single element can fulfil several functions thus leading to simplified and economical construction. These functions can include structure, fire protection, sound insulation, thermal insulation, weather exclusion and sub-division of space. Masonry materials are available with properties capable of meeting these varied requirements, requiring only to be supplemented by the use of other materials for thermal insulation, damp-proof courses, wall ties and the like.

The second major advantage relates to the durability of masonry materials which, with appropriate selection, can be expected to remain serviceable for many decades, if not centuries, with relatively little maintenance.

From the architectural point of view, masonry offers great flexibility in terms of plan form, spatial composition and facing for which materials are available in an almost infinite variety of colours and textures. These features offer great flexibility in design, as has been demonstrated in buildings of many different categories. It might be noted, however, that one architectural advantage of masonry construction has been exploited only to a limited extent in recent times, namely, the ease with which curved walls can be built, economically and without complication, thus enabling the architect to depart from the rigidity of the rectangular grid which is so much a feature of modern building. The architectural possibilities are also extended by the use of reinforcement or prestressing.

The nature of masonry is such that its construction can be achieved without very heavy and expensive plant, a factor that leads to economy. Although dependent on skilled labour for a high standard of construction, productivity has been maintained by the use of larger units, improved materials handling and off-site preparation of mortar. Further development of site methods has taken place in recent years, particularly in Germany and the Netherlands and there are at least limited possibilities for prefabrication all of which can be expected to maintain the competitive position of masonry construction.

The advantages of masonry wall construction are thus considerable but, as with all materials, appropriateness to the application has to be considered. Thus both overall and detail design must have regard to the characteristics of the material. Most problems encountered in masonry buildings arise from differential movement between dissimilar materials and from rain penetration. Avoidance of such defects depends on the adoption of suitable materials and details and the achievement of a reasonable standard of construction. Foundation movement is also a source of trouble but is not specifically a problem of masonry construction.

1.4 Factors affecting the design of masonry buildings

Before proceeding to the design of masonry walls as such, it will be necessary to give attention to a number of factors relating to the building as a whole.

Thus it is necessary to consider the implications of the weight of the masonry as it affects the supporting structure if the masonry is not loadbearing. If the structure is loadbearing it is important to ensure that the layout of the walls is consistent with overall stability and is such as to avoid susceptibility to failure in the event of accidental damage. It must also be considered whether the area taken up by masonry walls is significant in relation to the available floor area.

From the construction point of view, availability of the necessary skilled labour, the construction time and its phasing with the overall building schedule will also be relevant factors at the preliminary design stage.

Having resolved these questions, the masonry will be designed to meet criteria relating to imposed loading, thermal, acoustic and fire conditions and resistance to rain penetration. Consideration must also be given to durability and movement, particularly in relation to contiguous elements or materials. Appearance of external surfaces in terms of colour and texture is important and if applied facings or finishes are required, compatibility with the masonry has to be assured.

The various design parameters are specified in codes of practice. Thus for example, loading requirements for buildings are set out in Eurocode 1 (EC 1) for dead, imposed, fire, snow and wind loads. This document refers to 'actions' which may be either 'direct', i.e. a force or load, or 'indirect' which may result, amongst other things, from an imposed or constrained deformation. Other requirements may be imposed by building regulations: for example, in the UK measures to protect the public against the consequences of failure resulting from accidental damage are laid down in such regulations. Eurocode 6: Part 1-1 sets out the basis for structural design of masonry walls and Part 2 general design, selection of materials and execution. Effective use of these documents is, however, dependent on a full understanding of the properties of the materials being used, the relevant structural principles and construction methods which will be discussed in subsequent chapters of this book.

1.5 Future trends

As indicated above, the predominant use of masonry walling is in low to medium rise buildings and for cladding of steel and concrete frame structures. In the UK the standard construction for external walls is the cavity wall, typically with clay brick outer leaf and concrete block inner leaf. The choice of materials for such walls is increasingly being determined by thermal insulation requirements. In many parts of Europe where driving rain is less severe there is a preference for solid walls built from thicker blocks. Again, however, thermal requirements are an increasingly important factor driven by the need for higher energy efficiency.

For masonry to retain its place as a primary construction material, particularly as a cladding, there will be a need to improve building techniques with a view to reducing construction time on site. The means of achieving this may be expected to include the use of larger units requiring the use of mechanical handling in placing them on the wall. Such units will be dimensionally accurate leading to thinner jointing and the use of smaller quantities of mortar, possibly of new types.

In the Netherlands, a method has been developed whereby units for a complete wall are delivered to site rather than simply by consignment of units, some of which may have to be cut. In this connection site time may be saved by modular dimensioning of walls having regard to standard unit sizes. Prefabrication of walls has been attempted in the past but has considerable logistic and economic limitations. Future use of off-site construction is likely to be limited to smaller elements, which would not call for highly expensive factory plant, transportation and lifting equipment.

In addition to these well known trends there has recently entered into prominence the need to take account of 'sustainability' and possible climate change in decisions relating to building design, particularly in response to the now generally accepted phenomenon of global warming. This has two aspects: first, whilst the causes and mechanism of climate change are not fully understood, it is undoubtedly exacerbated by the release of carbon dioxide into the atmosphere so that in seeking to reduce such emissions there is a need to adopt methods of construction and use of buildings which will have this effect. Secondly, although global warming may result in an apparently rather small mean temperature increase, it may result in weather changes leading to greater extremes being experienced. Thus in some places there may be a greater frequency of storm force winds, in others higher rainfall or the reverse, an increase in summer temperatures with consequent liability to desiccation of clay soils. Also, a rise in sea level is expected, leading to flooding in many low-lying areas. All of these effects have implications for building performance and construction. Although their extent is uncertain and may not be evident for a considerable time, buildings are expected to last for many decades and therefore prudence dictates that due allowance in the design of buildings be made in the immediate future for the possible effects of global warming.

The particular implications for the use of masonry may include further requirements for improved thermal efficiency, both for heating and cooling, to reduce the need for CO_2 producing fuels. The same pressure will lead to preference being given to masonry units, which have relatively low energy demand in their production. On the whole, however, it can be expected that masonry will be well placed to meet sustainability requirements.

Further reading

Lynch, G. (1994) *Brickwork: History Technology and Practice*, Donhead, London.
The role of brickwork in our environment, Hammett, M., Brick Development Association, BDA Publ. SP 2, 1991.
The design of curved brickwork, Hammett, M. and Morton, J., Brick Development Association, BDA Publ. DN12, 1991.

Chapter 2

Masonry units: bricks, blocks and natural stone

2.1 Masonry units: general

Masonry walling units are produced from clay, calcium silicate and concrete or shaped from stone extracted from quarries. Joining these units together by cement or lime mortar or other binding material produces external and internal masonry walls of different kinds for building and civil engineering structures. All walling units have broadly similar uses although their properties differ in some important respects depending on the raw materials used and the method of manufacture. The manufacturing processes are largely mechanised today although clay bricks and natural stone blocks are still made or shaped by hand in many parts of the world.

Bricks and blocks are produced in many forms as summarised in Table 2.1 and illustrated in Fig. 2.1. The size of bricks varies from one country to another but typically is as shown in Table 2.2. The 'coordinating size' is that of the 'work size' of the brick plus the nominal thickness of the mortar joint, usually 10 mm. The coordinating size is used in dimensioning walls whilst the actual size is based on the work size making allowance for shrinkage occurring during manufacture.

Blocks are larger than bricks and are produced in a considerable range of sizes as indicated in Table 2.3.

2.2 Manufacture of units

2.2.1 Clay bricks and blocks

Clay bricks are manufactured from clay, clayey soils or soft slate or shale. The best material for brick making is clay containing about 30% sand, which reduces the shrinkage occurring during the burning of soft clay. Clays are first ground or crushed in mills and mixed with water, as required to achieve the desired consistence for forming.

There are essentially three methods of forming known as soft mud, dry press and wire cut. In the soft mud process preparation of the clay may include the addition of crushed reject bricks, lime or pulverised fuel ash and organic matter to act as fuel. After mixing and blending the clay is hydraulically pressed in steel moulds the faces of which may be sanded or wetted to prevent sticking of the clay. Clearly, solid bricks result from this process but they may have depressions or frogs on one or both of the bed faces. The dry press process is similar but uses a stiffer clay mix with lower water

Table 2.1 Types of masonry units

Type of unit	Description
Bricks	
Solid	A brick which may have perforations not exceeding 25% of its volume or may have indentations (frogs) on one or both bed faces
Perforated	A brick which has a pattern of small holes through it comprising more than 25% of its volume
Cellular	A brick in which the holes comprising more than 20% of its volume are closed on one face
Hollow	A brick having large holes through it comprising more than 25% of its volume
Blocks	
Solid	A block without holes or cavities
Cellular	A block having one or more cavities which do not pass through the block
Hollow	A block having one or more cavities which pass through the block

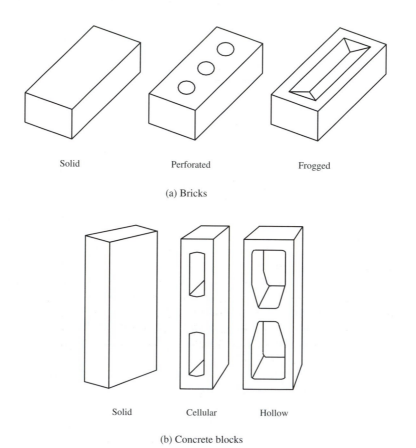

Solid   Perforated   Frogged

(a) Bricks

Solid   Cellular   Hollow

(b) Concrete blocks

Figure 2.1 Common types of bricks and blocks.

Table 2.2 Sizes of bricks

Coordinating size			Work size		
Length (mm)	Width (mm)	Height (mm)	Length (mm)	Width (mm)	Height (mm)
225	112.5	75	215	102.5	65

Table 2.3 Sizes of concrete blocks

Work size + 10 mm	Specification dimensions (mm)	
Coordinating size (Length × height)	Work size (Length × height)	Work size (Thickness)
400 × 200	390 × 190	60, 75, 90, 100, 115, 140, 150, 190, 200
450 × 150	440 × 140	60, 75, 90, 100, 140, 150, 190, 200, 225
450 × 200	440 × 190	60, 75, 90, 100, 140, 150, 190, 215, 220
450 × 225	440 × 215	60, 75, 90, 100, 115, 125, 140, 150, 175, 190, 200, 215, 220, 225, 250
450 × 300	440 × 290	60, 75, 90, 100, 140, 150, 190, 200, 215
600 × 150	590 × 140	75, 90, 100, 140, 150, 190, 200, 215
600 × 200	590 × 190	75, 90, 100, 140, 150, 190, 200, 215
600 × 225	590 × 215	75, 90, 100, 125, 140, 150, 175, 200, 215, 225, 250

content. This results in accurately formed bricks but is more expensive and therefore not so widely used.

In the wire cut method a stiff clay mix containing 20–25% water is forced through a die to form a ribbon of the desired format and size of the brick or block. The continuous column of clay is then cut into units by wires set apart by the height of the brick or block plus an allowance for shrinkage. Most structural clay units are manufactured by this process and this results in hard and dense products. Using this method, it is easy to include holes and perforations along the length of the column by placing appropriately shaped blockages in the die. The inclusion of such holes or perforations reduces the weight of clay at every stage of the production process and thus all clay preparation costs: shredding, grinding, mixing etc. Reduced use of clay also has environmental impact by reducing the rate of use of clay deposits and by improving the efficiency of fuel use in firing. Finally, perforated units have improved thermal characteristics and are less tiring to lay as a result of their lighter weight.

In some cases the surface of the formed unit is mechanically treated to give a texture when eventually exposed in facing brickwork.

After forming, the units have to be dried at temperatures between 40 and 150°C for 24–42 hours the heat for this being recovered from the firing kiln. Modern brick plants use tunnel kilns in which the green bricks are stacked on cars which move through the kiln exposing the bricks first to gradually increasing then to decreasing temperatures. The maximum firing temperature depends on the type of clay and on the desired characteristics of the bricks. The majority of clays burn to a red colour when fired at 900–1000°C. Above this temperature, the colour turns to dark red or purple then to brown or grey at 1300°C. The colours, however, depend on the mineral content of the clay, thus a high iron content makes the brick salmon pink when fired at 900°C whilst lime or chalk may result in a white or cream colour.

2.2.2 Calcium silicate bricks and blocks

Calcium silicate units are manufactured from lime, silica sand or crushed siliceous rock with colouring additives. A variation uses crushed flint in place of silica sand. These materials are mixed, allowing sufficient time for complete hydration of the lime before forming the units by moulding under pressure. The bricks or blocks are then cured by steam pressure in an autoclave for several hours, hardening the material by chemical action.

Calcium silicate units have accurate dimensions, allowing for some shrinkage after manufacture, with characteristically sharp arrises requiring care in handling, as damage is very apparent in the finished masonry. In appearance they are very uniform in colour of which a wide variety is available. White to pink and other pastel shades are most commonly used but the colour darkens appreciably when the surface is wet.

Calcium silicate bricks are used in the UK for general and facing purposes but to a much lesser extent than clay bricks. In some European countries blocks of various sizes and types in this material are widely used.

2.2.3 Concrete bricks and blocks

Concrete bricks and blocks are produced in all the types included in Table 2.1. The basic material used is a dry, small aggregate concrete mix carefully controlled in a centrally located plant. Two main processes are used to mould the units differing essentially in the method of curing. The first is known as the 'egglayer' in which the mix is introduced into a steel mould, vibrated and compacted and the resulting block deposited on a concrete curing apron. The egglayer machine then moves on by steps to place further blocks on the apron. The second method uses a 'static' machine which feeds the formed blocks on to pallets thence by conveyor belt and transfer car to a curing chamber where they are usually treated with low pressure steam.

As indicated in Table 2.3 concrete units are produced in a wide variety of types and sizes and with external face treatments to enhance appearance of the wall. Special blocks are also made to simulate natural stone.

2.3 Properties of bricks and blocks

2.3.1 Physical properties

The following physical properties of masonry units are relevant to their use in the construction of walls:

1. Colour.
2. Surface texture.
3. Density and weight.
4. Absorption and pore structure.

5.   Thermal conductivity.
6.   Thermal and moisture movement.
7.   Fire resistance.

The range of products available is so large that it is not possible to give more than an indication of the characteristics of units in the materials discussed in preceding paragraphs.

As already suggested, clay bricks are produced in a variety of colours depending on the mineral content and firing temperature but most commonly in shades of red. However, in current architectural practice yellow, buff and brown facing bricks are frequently chosen, very often with a roughened surface finish. Combinations of different colours in the same wall are also used to add interest to the appearance of a building.

Calcium silicate and concrete bricks are also available in a range of colours, usually light grey and other paler shades. These bricks tend to give a more uniform appearance to a wall than clay bricks and look darker when wet. Concrete blocks are grey unless pigmented cement has been used in their manufacture. If enhanced appearance to exposed faces is required this can be achieved by painting, plastering or by the use of special blocks having a surface finish textured in one of a number of possible ways in the course of manufacture e.g. by tooling the surface or by exposing the aggregate.

The density of clay, calcium silicate and concrete is around $2 \, \text{tonnes/m}^3$ but the weight of units which is of more immediate practical importance depends on their size and type i.e. whether solid, cellular or perforated. Various lightweight materials, including in particular aerated autoclaved concrete with a material density in the range $450-850 \, \text{kg/m}^3$, result in solid blocks of a given size proportionately lighter than those of dense materials. Apart from the structural and sound insulating implications, unit weight is of importance in relation to handling during construction, as will be discussed in a later chapter.

The absorption and pore structure of bricks and blocks varies widely and has a number of important implications for the behaviour of the units. Clay bricks may absorb between 4.5 and 7.0% of their weight, low absorption being important if they are to be used as a damp-proof course. Another property is the Initial Rate of Absorption (IRA) which in particular gives an indication of the amount of water that a brick will absorb when it is laid on the mortar bed. The IRA is expressed in $\text{kg/m}^2/\text{min}$ a value of up to 1.5 in these units being satisfactory: if it is very low it becomes difficult to lay the bricks and if too high the removal of water from the mortar may prevent complete hydration of the cement. Absorption is of less relevance in the case of calcium silicate and concrete units but pore structure affects resistance to frost damage, as will be discussed in relation to durability in Chapter 5.

Thermal conductivity of units in the materials under consideration is of considerable practical importance. There is not a great deal of difference between solid units in clay, calcium silicate and dense concrete but lightweight aggregate and AAC blocks have substantially lower thermal conductivity than the heavier materials. Thermal conductivity, measured in units of watts/metre thickness per degree Kelvin (W/mK), is 0.84 for clay bricks as compared to 0.11 for lightweight blocks. Hollow and perforated clay and concrete units will have intermediate values depending on their

Table 2.4 Thermal and moisture movement in masonry units

Material	Coefficient of linear expansion $\times 10^{-6}$	Moisture movement (%)	
		Irreversible	Reversible
Clay bricks	4–8	+0.1–0.2	Negligible
Concrete	7–14	−0.02–0.06	0.01–0.06
Calcium silicate	11–15	−0.01–0.04	0.01–0.05

characteristics. The insulating properties of complete walls, however, depend on a number of factors in addition to the thermal properties of the units (cf. Chapter 5).

Thermal and moisture movement of masonry walls requires to be taken into account in design. An indication of these movements in various units is shown in Table 2.4 but values for masonry are more useful and are given in Chapter 5.

Fire resistance of masonry units is of practical importance but as the materials are inherently resistant, the critical factor in this respect lies in the detail design of the construction aimed at preventing fire passing through defects in, or finding a way around, a wall.

2.3.2 Mechanical properties

The most important mechanical property of masonry units is compressive strength which, as well as being of direct relevance to the strength of a wall, serves as a general index to the characteristics of the unit. It is measured by a standardised test, the result depending to some extent on the conditions prescribed in the particular standard being used. It is important to note also that the apparent compressive strength obtained depends on the dimensions and type of unit. Thus if a brick and a block of larger overall dimensions but of the same material were tested, a higher figure would be obtained for the brick. This is because the material under compression deforms laterally and this deformation is resisted by the platens of the testing machine leading to an increase in the apparent strength of the material. The effect is greater in the brick which thus appears to have a higher strength per unit of area.

Clay bricks are obtainable in strengths of up to $100 \, \text{N/mm}^2$ but much lower strengths—say $20–40 \, \text{N/mm}^2$—are normally sufficient for domestic building and cladding for taller buildings. Concrete blocks have a lower range of apparent compressive strengths but if brickwork and blockwork are built of units of the same apparent compressive strength, the latter will have a greater strength in compression than the former. This is because dilation of the mortar joints has the reverse effect to platen restraint in the testing machine and tends to cause tensile failure of the units. As the joints are of similar thickness in both cases, the greater depth of the blocks will result in the paradoxical result of a higher masonry strength being obtained from the seemingly lower strength material.

The tensile strength of masonry units—both direct and flexural—has an influence on the resistance of masonry under various stress conditions but is not normally specified or measured except in relation to concrete partition blocks where a breaking strength of $0.05 \, \text{N/mm}^2$ is required.

2.4   Natural stone

2.4.1   General

Prior to the 20th century, natural stone was the predominant material used in major building construction. It is still used to a limited extent not only as a structural material but also for exterior cladding and interior finish of walls and for other purposes in building. Limitations on the use of natural stone arise from the massive weight of stone walling and the resulting foundation requirements as well as the considerable skilled labour requirement in preparation and laying of the stones. Nevertheless, appearance and durability of stone masonry are valuable advantages and there has been an increase in the use of the material for cladding in recent years. Furthermore, it is likely that considerations of environment and sustainability will encourage the use of stone masonry in future.

Despite their great variety, relatively few types of stone are suitable for building masonry walls. In addition to accessibility and ease of quarrying, the stone must satisfy the requirements of strength, hardness, workability, porosity, durability and appearance. Some of the stones that satisfy these requirements are granite, limestone and sandstone. Marble and slate are used for special purposes and others such as quartzite and serpentine are used locally or regionally, but to a much lesser extent.

2.4.2   Granite

Granite is an acid igneous rock which has a fairly limited range of composition. The granite family includes members which contain varying proportions of quartz, alkali plagioclase feldspar, potash feldspar and a dark-coloured ferromagnesian mineral, generally either biotite mica or hornblende. Colours vary depending on the amount and type of secondary minerals. The mineral present in the greatest quantity is feldspar. Feldspar produces red, pink, brown, buff, grey and cream colours, while hornblende and mica produce dark green or black. Quartz is usually found to be colourless, however, in the main mass of the rock, it appears to be grey, although it may sometimes have a pale purple hue. The dark brown mica, biotite, is evenly scattered throughout the mass and, if the pale-coloured mica, muscovite, is present, it is distinguished by a silvery appearance.

Granites are well known for their durability in many types of environment. Granite has been used as a building material almost since the inception of man-made structures. Because of its hardness, it was first used with exposed, hand-split faces. As tools and implements became more refined, the shapes of the stone also became more sophisticated. With the development of modern technology and improved methods of sawing, finishing and polishing, granite was more readily available in the construction market and more competitive with the cost of other, softer stones.

Granite is classified as fine, medium or coarse grained. It is very hard, strong, durable and is noted for its hardwearing qualities. Compressive strength may range from 50 to 415 N/mm^2. While the hardness of the stone lends itself to a highly polished surface, it also makes sawing and cutting very difficult. Granite is often used for flooring, panelling, veneer, column facings, stair treads and flagstones or in landscape applications. Carving or lettering on granite, which was formerly done by hand or pneumatic tools, is now done by sandblasting, and can achieve a high degree of precision.

2.4.3 *Limestone*

Limestone is a sedimentary rock which is durable, easily worked and widely distributed throughout the earth's crust. Limestones consist chiefly of calcium carbonate ($CaCO_3$) deposited by chemical precipitation or by the accumulation of shells and other calcareous remnants of animals and plants in sea water and may be made up of as much as 99% calcium carbonate. They are commonly highly fossiliferous. Other mineral matter may be present in significant amounts. When pure, or nearly so, limestones are white in colour. Few limestones, however, are composed entirely of calcium carbonate. Most contain other mineral matter which will determine its overall colour. Clay is one of the commonest non-calcareous constituents of limestone. Limestones are very widespread, particularly in the Midlands and south of the UK.

Limestones can be classified into three types on the basis of their natural formation and origin. Chemical limestones are formed directly by precipitation of calcium carbonate from water. Organic limestones consist largely or entirely of the fossilized shells of one or more organisms. The live organisms removed calcium carbonate from the water in which they lived and used it for their shells or skeletons. Clastic or detrital limestones result from the erosion of pre-existing limestones. Many limestones contain magnesium carbonates in varying proportions, sand or clay, carbonaceous matter, or iron oxides, which may colour the stone. The most 'pure' form is crystalline limestone, in which calcium carbonate crystals predominate, producing a fairly uniform white or light grey stone of smooth texture. It is highest in strength and lowest in absorption of the various types of limestone. Dolomitic limestone contains between 10 and 45% magnesium carbonate, is somewhat crystalline in form, and has a greater variety of texture. Oolitic limestone consists largely of small, spherical calcium carbonate grains cemented together with calcite from shells, shell fragments, and the skeletons of other marine organisms. It is distinctly non-crystalline in character, has no cleavage planes and is very uniform in composition and structure.

The compressive strength of limestone varies from 20 to 190 N/mm^2 depending on the silica content, and the stone has approximately the same strength in all directions. Limestone is much softer, more porous and has a higher absorption capacity than granite, but is a very attractive and widely used building stone. Although stone may look solid on first sight, it does in fact contain many pores. Some of the pores are large enough to see with the naked eye but most can be seen only under a microscope. Limestones are soft when first taken from the ground but they weather hard upon exposure. The impurities affect its colour, iron oxides produce reddish or yellowish tones as pink, buff or cream, while organic materials such as peat give a grey tint. Limestone textures are graded as statuary, select, standard, rustic, variegated and old Gothic. Statuary, select, standard and rustic come in buff or grey, and vary in grain from fine to coarse. Variegated is a mixture of buff and grey, and is of unselected grain size, whereas Gothic is a mixture of rustic and variegated and includes stone with seams and markings.

Limestone is used as cut stone for veneer, caps, lintels, copings, sills and ashlar with either rough or finished faces. Roughly dressed or quarry stone is often used as a rustic veneer on residential and low-rise commercial buildings. When the stone is set or laid with the grain running horizontally, it is said to be on its natural bed. When the grain is oriented vertically, it is said to be on edge but is generally to be avoided as it increases liability to frost damage.

2.4.4 Sandstone

Sandstone is a sedimentary rock formed of sand or quartz grains cemented together by matrices of different compositions. All these cementing materials were carried into the original loose, incoherent sand by circulating waters or by wind. The most common mineral grains are quartz, micas, feldspars and clays. The cementing medium holds the grain together but does not necessarily completely fill the voids between the grains. Porosity will be found both between and within grains. Its hardness and durability depend primarily on the type of cementing agent present. If cemented with silica and hardened under pressure, the stone is light in colour and very strong and durable. If the cementing medium is largely iron oxide or iron hydrates, the stone is light red or deep brown, and is softer and more easily cut.

Dolomitic sandstones are those whose grains are cemented with dolomite. These tend to be more resistant to acid-based weathering, compared with calcareous sandstones. Calcareous sandstones are sandstones whose cementing component is calcite (calcium carbonate). Calcite is susceptible to attack from airborne acids and calcareous sandstones are therefore prone to deterioration in urban and industrial environments.

Argillaceous sandstones contain significant amounts of clay in their binding media. Argillaceous sandstones have very low durability and they are subject to disintegration by natural weathering. When first taken from the ground, sandstone contains large quantities of water, which make it easy to cut. When the moisture evaporates, the stone becomes considerably harder.

Sandstones vary in colour from buff, pink and crimson to greenish brown, cream and blue-grey. Both fine and coarse textures are found, some of which are highly porous and therefore low in durability. The structure of sandstone lends itself to textured finishes and to cutting and tooling for ashlar and dimension stone in veneers, mouldings, sills, and copings. Roughly shaped sandstone is also used in rubble known in some areas as ragstone.

Further reading

Bricks: note on their properties, Hammett, M., Brick Development Association, BDA Publ. TIP 7.

Perforated clay brick, Building Research Establishment, BRE Digest 273, 1983.

Calcium silicate brickwork, Building Research Establishment, BRE Digest 157, 1992.

Autoclaved aerated concrete, Building Research Establishment, BRE Digest 342, 1989.

The selection of natural building stone, Building Research Establishment, BRE Digest 269, 1983.

Selecting natural building stones, Building Research Establishment, BRE Digest 420, 1997.

Clay bricks, British Standards Institution, BS 3921: 1985.

Precast concrete masonry units, British Standards Institution, BS 6073: Part 1: 1981.

Specification for calcium silicate (sandlime and flintlime) bricks, British Standards Institution, BS 187: 1978.

Chapter 3

Mortar and other components

3.1 Mortar

3.1.1 General

Although mortar accounts for as little as 7% of the total volume of masonry, its influence on the masonry assemblage is far more than this proportion indicates. The primary function of mortar is not only to bind the individual units together so that the masonry will act as larger single unit elements, but also to seal against air and moisture penetration. It bonds with steel reinforcement, metal ties and anchor bolts to join building components together and as a binding material, the mortar is partly responsible for the strength characteristics of the masonry. Mortar joints also contribute to aesthetic features such as colour and texture.

The most important physical properties of freshly made mortar include workability, cohesiveness and the ability to spread evenly and easily. It should not lose water readily and stiffen on contact with absorptive masonry units but should remain plastic long enough for bricks or blocks to be adjusted to the desired line and level. Stiffer mortars, however, are needed for non-absorptive units. The critical characteristics of the hardened mortar are bond strength, resistance to rain penetration, durability and compressive strength.

Workability significantly influences most other mortar characteristics. A workable mortar has a smooth, plastic consistency, is easily spread with a trowel and easily adheres to vertical surfaces. Well-graded, smooth aggregates enhance workability as do lime, air entrainment agents and correct amounts of mixing water. Lime imparts plasticity and ability to retain water known as 'retentivity'. Air entrainment improves the frost resistance of both freshly laid and hardened mortars. Mortar in external masonry must be frost resistant when it has hardened. This is achieved by batching and mixing mortar correctly and to the appropriate specifications for the degree of exposure. Hardened mortar may need to resist the effects of soluble salts in particular sulphates present in some types of units or arising from ground water or the atmosphere.

Durability and load bearing ability require that mortar develop early strength as it hardens to allow building to proceed without unnecessary delay. However, a final compressive strength of $2-5\,\text{N/mm}^2$ when mortar is fully cured is adequate for most low-rise masonry structures. Mortar within this strength range will have the ability to accommodate small movements and any cracking in the masonry will usually be distributed as hair cracks in the joints where they are not easily seen and do not influence the wall stability. However, weaker mortars will not be durable under severe conditions but using unnecessarily strong mortars will concentrate the effects of movement in fewer and wider cracks in the wall unless adequate movement joints are provided.

3.1.2 Cements for mortar

The most commonly used cement in mortar is ordinary Portland cement but rapid hardening, white, blast-furnace and pozzolana cements are also used in particular circumstances. Certain cements, such as masonry cement, contain additives that impart special properties such as delayed or accelerated rates of setting, resistance to chemical attack, enhanced workability or colour.

(a) Ordinary Portland cement

Ordinary Portland cement is the principal binding ingredient used in most mortar mixes currently used. It is made primarily from a combination of calcareous material, such as limestone or chalk, and of silica and alumina found as clay or shale. The process of manufacture consists essentially of grinding the raw materials into a very fine powder and mixing intimately in predetermined proportions, with or without water. The mixture is then fed into a rotary kiln and burned at a temperature of about 1400°C when the material sinters and partially fuses into clinker. The clinker is cooled and ground to a fine powder with some gypsum added.

The addition of water to Portland cement powder results in a plastic paste and starts a chemical reaction. The paste begins to set within a few hours and sets finally within about 10 hours following which hardening continues for a long period of time. However, 80–90% of the final strength is reached in the first month as a result of the hydration of silicates and aluminates contained in the cement.

Ordinary Portland cement is by far the least expensive and most widely used type of cement and is suitable where there is no exposure to sulphates from masonry units or from groundwater.

(b) Rapid hardening cement

Rapid hardening cement is similar in manufacture to ordinary Portland cement but is ground more finely. As the name implies, this cement has higher early strength developing similar strength at 5 days as ordinary Portland cement at 7 days, given the same water/cement ratio, but both cements result in approximately the same final strength. The increased rate of hydration is accompanied by a higher rate of heat development, making it unsuitable for large masses of concrete. On the other hand, this may be an advantage in construction at low temperatures as a safeguard against frost damage.

The cost of rapid hardening Portland cement is only marginally higher than that of the ordinary variety.

(c) White and coloured Portland cements

These types of cements are similar in basic properties to ordinary Portland cement; their uses are generally for aesthetic reasons where a white or other colour rather than the normal grey is required. White cement is made from china clay together with chalk or limestone free from specified impurities. White cement may be interground with a pigment to produce other colours or a similar effect can be obtained by adding pigments to the mixer, provided that there is no adverse effect on strength.

White Portland cement is more expensive than ordinary Portland cement as precautions are necessary in manufacture to avoid discolouration by iron and other contaminants.

(d)   Sulphate resisting Portland cement

This type of cement is low in calcium aluminate (C_3A) giving it a high resistance to sulphate attack. This compound reacts with sulphates in solution to form calcium sulpho-aluminate and gypsum, which have increased volume in a hardened mortar resulting in its gradual disintegration.

Sulphate resisting cement is more expensive than ordinary Portland cement due to the special composition of the raw materials and is thus only specified when conditions make it necessary.

(e)   Portland-pozzolana cements

These types of cements are based on ordinary Portland cement to which has been added natural or artificial pozzolanic materials which react with calcium hydroxide, liberated by the hydrating cement, to form compounds possessing cementitious properties. Such materials include volcanic ash, pumice, fired clay and pulverised fuel ash (pfa).

The rate of development of strength of these cements depends on the activity of the pozzolana and on the percentage of cement in the mixture; it is generally lower than that of ordinary Portland cement as is the liberation of heat. The advantage of this type of cement lies in their high resistance to chemical attack.

(f)   Masonry cement

Masonry cement typically contains approximately 75% of ordinary Portland cement together with a fine ground mineral filler and an air-entraining agent. When mixed with sand and water, masonry cement results in a mortar with very good working properties and strength.

3.1.3   Sands for mortars

Mortar sands can originate from sand pits, sand dunes or have been dredged from rivers or the sea. Crushed brick, stone, clinker or slag may also be used. Of considerable importance in relation to the characteristics of the mortar resulting from the use of a particular sand is its particle size grading. Thus British standards define types S and G sands on this basis (as in Table 3.1) the latter comprising a considerably higher percentage of fines or 'silt'. Whilst silt improves workability it tends to detract from durability in severe exposure conditions or may reduce flexural strength. Coarser sands will be preferable in these conditions although not easily worked. Clay also acts as a plasticiser but too high a clay content will retard the set of the mortar and

Table 3.1 Sands for mortar for plain and reinforced brickwork, blockwork and masonry (after BS 1200)

BS Sieve (mm)	Percentage by mass passing BS sieves	
	Type S	Type G
6.30	100	100
5.00	98 – 100	98 – 100
2.36	90 – 100	90 – 100
1.18	70 – 100	70 – 100
0.60	40 – 100	40 – 100
0.30	5 – 70	20 – 90
0.15	0 – 15	0 – 25
0.075	0 – 5	0 – 8

increase the water content for a workable mix resulting in reduced strength and increased shrinkage. Ideally, the particle size distribution should be such that the smaller fractions should occupy the gaps between the larger grains with the cement and lime filling the remaining small spaces.

Sands may contain a variety of impurities including organic and mineral matter. Iron pyrites, coal and shale are potentially harmful as are humus and industrial waste. Sea sands will contain some residual salt which is unlikely to affect the set of the mortar but may lead to corrosion of embedded steel or other metal as well as to dampness and efflorescence.

3.1.4   Water for mortars

It is essential to use water free from contaminants for mixing mortar as the presence of impurities may have an adverse effect on the properties of the hardened material. Drinking water will be of a satisfactory standard and, if not available, the quality of a proposed source may have to be checked. In particular, sea water with its high salt content should not on any account be used.

3.1.5   Admixtures used in mortar

An admixture is a material added to fresh mortar to change one or more of the properties of the fresh or hardened mix. The term 'admixture' is used in preference to 'additive' which is used by cement manufacturers to identify substances incorporated in cement to modify their behaviour e.g. gypsum to slow down flash setting. Most admixtures are added to mortar in very small, carefully controlled quantities. The most commonly used admixtures are accelerators and frost inhibitors, retarders, air entraining agents, plasticisers and pigments.

3.1.6   Mortar mix design

Since the late 19th century, Portland cement has become the major binding ingredient of masonry mortar. Occasionally, it is used with sand and water in what is called a straight cement mortar. Such a mix in proportions 1 cement: 3 parts sand by volume gives a mortar which hardens quickly and consistently, exhibits high strength and good resistance to frost. However, it has poor workability, low water retentivity and poor bond. The use of such cement mortar is therefore restricted to special situations where its favourable properties over-ride its disadvantages. A range of generally useful mortars is produced by replacing part of the cement in a 1 : 3 cement : sand mix by an equal volume of lime so that the binder paste still fills the voids in the sand. In this type of mix, the cement contributes durability, high early strength, a consistent hardening rate and high compressive strength; lime adds workability, water retention and bonding properties and elasticity. Cement : lime mixes usually produce highly satisfactory mortars with a good overall performance.

In current British practice there are five mortars of decreasing strength but increasing ability to accommodate movement due to temperature and moisture change. These are shown in Table 3.2 for cement : lime : sand mixes ranging from $1 : 0 - \frac{1}{4} : 3$ to $1 : 3 : 10 - 12$ proportions by volume and from 11 to $1.0 \, \text{N/mm}^2$ in compressive strength. Designation (iii) is referred to as 'General purpose mortar' and, as this description implies, is regarded as suitable for most construction. However, in highly stressed brickwork designation (i) may be appropriate whilst designation (iv) will be suitable for relatively low strength blockwork. Exceptionally, different mixes may be specified for the inner and outer leaves of a cavity wall where these are of materials having unlike properties and the exposure conditions are obviously different. However, it is desirable in such cases to use the 1 : 1 : 6 mix throughout to avoid confusion on site.

Whilst cement : lime : sand and cement : sand with plasticiser are the most commonly used mixes, masonry cement and other pre-mixed products are sometimes used in the interests of accurate gauging and favourable working properties. Lime : sand mortar has been displaced by mixes containing cement because of their more rapid setting time resulting in higher rates of construction and the ability to work in all but the most severe weather conditions. Lime mortar, however, is preferred in restoration work as it will be consistent with the mortar in the building, will not cause damage to stone masonry and is able to accommodate considerable movement without visible cracking.

Other types of mortar may be specified, for example to possess maximum durability and resistance to water penetration or to meet other special requirements. The architect or engineer is best placed to select the type of mortar to be used in a particular case being in a position to identify and assess all the relevant design criteria and to ensure the quality of the construction through adherence to specification. The mason or mason contractor is of course responsible for achieving the required standard of workmanship but they inevitably have a prime interest in workability, rate of construction and general economy on site.

3.2 Components

In addition to units and mortar, masonry wall construction requires the use of a number of subsidiary components including damp-proof course material, cavity trays, wall ties and fixings. All of these must be as durable as the masonry itself as well as meeting its particular function.

Table 3.2 Requirements for mortar (after BS 5628: Part I)

Mortar designation	Proportions by volume of dry materials			Mean compressive strength of site-mixed mortars at 28 days (N/mm^2)
	Cement : lime : sand	Masonry cement : sand	Cement : sand with plasticiser	
(i)	$1:0-\frac{1}{4}:3$	—	—	11.0
(ii)	$1:\frac{1}{2}:4-4\frac{1}{2}$	$1:2\frac{1}{2}-3\frac{1}{2}$	$1:3-4$	4.5
(iii)	$1:1:5-6$	$1:4-5$	$1:5-6$	2.5
(iv)	$1:2:8-9$	$1:5\frac{1}{2}-6\frac{1}{2}$	$1:7-8$	1.0
(v)	$1:3:10-12$	—	—	—

Increasing strength.

Increasing ability to accommodate movement due to temperature and moisture changes.

Increase resistance to frost attack during construction.

Improvement in bond and consequent resistance to rain penetration.

Direction of change in properties is shown by arrows.

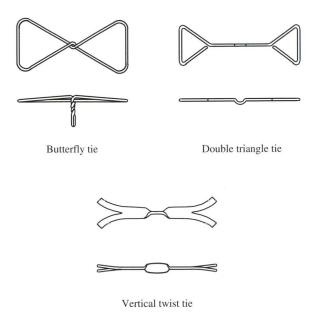

Butterfly tie   Double triangle tie

Vertical twist tie

Figure 3.1 Metal wall ties for cavity walls. (After BS 1243.)

3.2.1   *Damp-proof course material and cavity trays*

Damp-proof courses (DPCs) and cavity trays are incorporated in masonry walls to prevent the ingress of rain or sub-soil water to the building. Various types of material are suitable depending on the location and purpose of the component and include a range of bitumen composites, pitch/polymer and polythene products. These materials have the advantage of adaptability to many different applications and are the most commonly used. Sheet copper is suitable in highly stressed walls and lead is appropriate in restoration of old buildings but have to be coated with bitumen to prevent corrosion if in contact with mortar. Very dense engineering bricks or layers of slate can be used as a damp-proof course at the base of a wall if high compressive or bond strength is required.

Apart from cost, the main criteria for the selection of DPC material are ability to resist the expected compressive stress without excessive deformation and durability. On this basis, all the materials mentioned above are adequate for buildings up to four storeys in height but polythene or bitumen polymer, pitch polymer, copper or high strength brick would be required for higher buildings.

In some situations it is necessary to be able to form the material on site but many preformed cavity trays and roof flashings are available and are to be preferred for ease and accuracy of installation.

3.2.2   *Wall ties and fixings*

In cavity wall construction, the leaves have to be tied together with suitable ties. Several types are available as shown in Fig. 3.1 in galvanised or stainless steel.

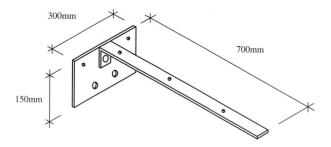

(a) Stainless or galvanised steel timber floor/masonry wall strap

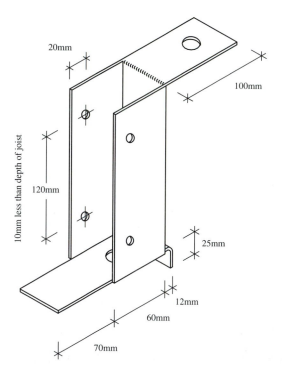

(b) Stainless or galvanised steel joist hanger

Figure 3.2 Timber floor/masonry fixings.

The latter are of course more expensive but are far more durable so that the extra cost, which is marginal in the cost of a wall, is fully justified in external walls in exposed locations. Conversely, galvanised butterfly ties are very susceptible to corrosion and their use is inadvisable in any but the most sheltered situations. Special wall ties used in repairing walls in which the ties have been incorrectly placed or omitted in construction or have become ineffective as a result of corrosion are described in Chapter 8.

The normal types of wall ties shown in Fig. 3.1 can accommodate relative move-ment between the leaves of a cavity wall over a height of three to four storeys but if the leaf height exceeds this, special ties are required. Such a need also arises when a masonry wall is used as the outer cladding to a concrete or steel frame as the move-ments of the two will be different. For this purpose dovetail ties and sliding fixings are used, as described in Chapter 5.

Different types of ties are required between brickwork cladding and timber frame walls. These are thin stainless steel, nailed to the timber and embedded in the mortar joints of the brickwork. Other galvanised and stainless steel straps or ties (Fig. 3.2(a)) are available for connecting timber floors to masonry walls to provide lateral support in two-storey buildings. Timber floor beams are frequently supported from masonry walls by joist hangers of the type shown in Fig. 3.2(b).

3.2.3 Reinforcement

Reinforced and prestressed masonry makes use of the same kinds of reinforcing bars and tendons as in the corresponding concrete construction for beams and other primary structural members. Various forms of bed joint reinforcement are used to increase the lateral resistance of thin wall panels.

Further reading

Building mortar, Building Research Establishment, BRE Digest 362, 1991.

A basic guide to brickwork mortars, Hammett, M., Brick Development Association, BDA Publ. No. 22, 1988.

Mortar and movement in aircrete blockwork: a review, Garvin S. L., Building Research Establishment, BRE Occasional Paper, 1994.

Aircrete: thin joint mortar building systems, Building Research Establishment, BRE Digest 432, 1998.

Damp-proof courses, Building Research Establishment, BRE Digest 380, 1993.

Performance specifications for wall ties, De Vekey, R. C., Building Research Establishment, BRE Publ. 45.

Ties for masonry cladding, Building Research Establishment, BRE Publ. IP 17/88, 1988.

Joist hangers, Building Research Establishment, BRE Publ. GBG 21, 1996.

Use of masonry: materials and components, British Standards Institution, BS 5628: Part 3: 1985.

Metal ties for cavity wall construction, British Standards Institution, BS 1243: 1978.

Chapter 4

Structural design

4.1   General considerations

In many low-rise masonry buildings design of the walls is determined by non-structural considerations such as resistance to rain penetration, thermal insulation or acoustic properties, as discussed in Chapter 5. Beyond two or three stories, however, specific attention has to be given to structural problems, beginning with the layout of the walls. This is primarily related to the function of the building but may have important structural implications and thus in a building of any size requires joint consideration by the architect and engineer.

In considering wall layout, it is necessary to ensure robustness of the building as a whole in terms of lateral strength and rigidity. In particular, the wall layout must be such that any local damage to the structure does not result in failure out of proportion to the initial cause. This is illustrated in Fig. 4.1 with reference to a simple cross-wall building. If this consisted only of floor slabs and walls as in Fig. 4.1(a) it would obviously be liable to collapse if subjected to small lateral forces. Stability would be improved by the addition of a service shaft (Fig. 4.1(b)) but damage to one of the cross-walls could still result in failure of a large part of the structure. A robust structure results from the provision of longitudinal return walls (Fig. 4.1(c)), which need not be load bearing. In this case removal of one of the cross-walls will not result in collapse of, or even significant damage to, the structure as a whole. Methods of protection against disproportionate failure as a result of accidental damage are given in the British Code of Practice, BS 5628: Part 1, for buildings of five stories and over but the general principles should also be observed in low rise construction.

Care should also be exercised in relation to unsymmetrical wall layouts which may result in undesirable distribution of forces between walls. This can occur in 'soft' ground floor layouts, for example where there is a garage at this level without adequate bracing walls. This has been known to result in severe damage to a low-rise building following a relatively small earthquake.

Robustness also depends on adequate tying together of the walls and floors. The latter should be capable of acting as a plate, distributing lateral forces to those walls capable of transmitting them to the foundations.

Other structural features which arise at the initial design stage include the incorporation of unusually long span floors, beams or lintels. The latter may give rise to heavy concentrations of load which could require the use of structural details that

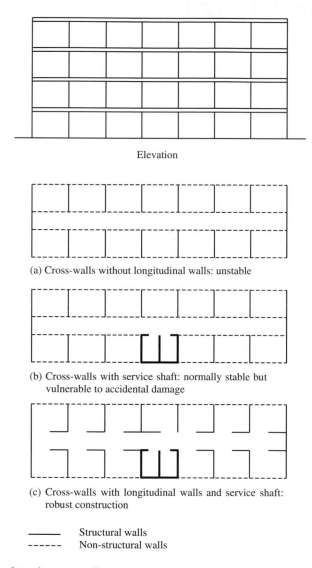

Elevation

(a) Cross-walls without longitudinal walls: unstable

(b) Cross-walls with service shaft: normally stable but
    vulnerable to accidental damage

(c) Cross-walls with longitudinal walls and service shaft:
    robust construction

——— Structural walls
------ Non-structural walls

Figure 4.1 Stability of simple cross-wall structure.

could affect the planning or appearance of the building. Similarly, the position of movement joints, if required, architectural features and means of accommodating services should be determined and taken into account before embarking on the structural design of the walls.

4.2 Loading on walls

Having decided the dimensions of the building and the wall layout, superimposed floor loadings appropriate to its use have to be determined and estimates made for

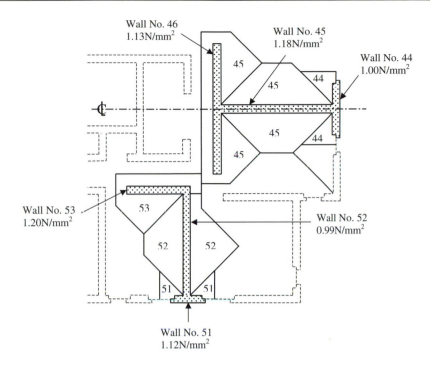

Figure 4.2 Wall loading by tributary areas.

self-weight of floors and walls. Such loadings are prescribed in standards such as Eurocode 1 and are multiplied by a partial safety factor to allow for unforeseen or abnormal events and other factors. Floor loads are then calculated and allocated to supporting walls. As the wall support forces generally form a statically indeterminate system, calculation of these forces is usually carried out on the basis of tributary areas the definition of which is to some extent a matter of judgment. This is illustrated in Fig. 4.2 in which the load from the floor area shown is supported by groups of two walls. The wall loads could reasonably be allocated by dividing the floor slab into a series of tributary areas or, alternatively, the total load could be concentrated at the centroids of the slab areas (Fig. 4.3). These will be eccentric to that of the wall groups and the wall loads are calculated by treating the walls as an I- or L-section under combined compression and bending.

This example assumes that all the walls are loadbearing and in cases where a slab crosses a non-loadbearing wall precautions have to be taken to ensure that there is no unintended transfer of load to such a wall by including a soft joint between the wall and slab.

The determination of wind loads on individual walls is more complex. First, it is necessary to establish the design wind pressure on the structure according to a prescribed standard, such as Eurocode 1: Part 1.4, which will take into account meteorological data for the site as well as the height and shape of the building. Thereafter, the wind loading acting on the face of external walls can be found.

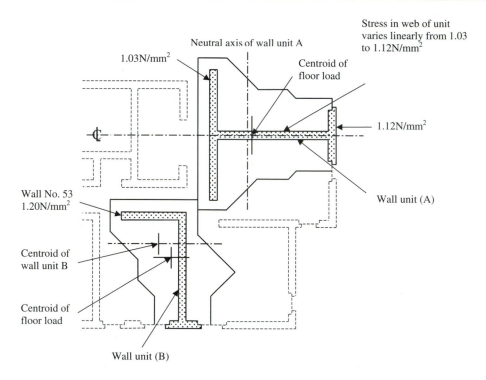

Figure 4.3 Wall loading concentrated at slab centroids.

The loading on walls which resist wind forces on the structure as a whole, is more complicated as such forces are distributed amongst the walls in proportion to their stiffness and on the nature of the interconnection between them. The simplest method of dealing with the problem is to assume that the walls act as cantilevers interconnected by pin-ended links as in Fig. 4.4 but this can be inaccurate, especially if the wall arrangement in plan is unsymmetrical. In the case of pairs of shear walls, however, this simple model is likely to err on the safe side so that if the result of such an analysis is satisfactory no more complicated investigation is required.

In some cases it may be necessary to consider stresses in walls resulting from the restraint of thermal movements but it is more usual for such effects to be avoided by provision of suitable expansion joints. In areas where earthquakes may occur appropriate account has to be taken in the design of masonry buildings. This is more realistically done by ensuring that the structure can accommodate a considerable degree of plastic deformation beyond its ultimate strength without collapse rather than attempting to design it to resist some arbitrary loading. For low level earthquake zones the measures taken to ensure general robustness and avoidance of severe accidental damage will be sufficient but if masonry buildings are to be built in high seismic areas more specific design measures with the adoption of reinforced walls will be essential.

Figure 4.4 Interconnected shear walls.

4.3   Design for compressive loading

The design of masonry walls in compression requires consideration of 'column effect' which in turn depends on the masonry strength and allowance for eccentricity of loading and slenderness of the wall.

4.3.1   Compressive strength of masonry

The compressive strength of the various types of masonry has been established on the basis of large numbers of tests on walls and prisms. It depends on the strength of the unit and on the mortar strength both of which are determined by standardised tests. In Europe, masonry strength is defined as a characteristic strength, in statistical terms a value which would be achieved by 95% of specimens tested. To avoid the need for numerous tables for the many varieties of masonry the idea of 'normalizing' the compressive strength of units to an equivalent 100 mm or 200 mm cube strength by the application of a shape factor has been developed. This has led to the following formula for characteristic strength:

$$f_k = K(f_b)^{0.65}(f_m)^{0.25} \tag{4.1}$$

where

f_k = characteristic compressive strength of masonry
K = factor depending on the type of masonry
f_b = 'normalized' compressive strength of the unit
f_m = mean compressive strength of the mortar

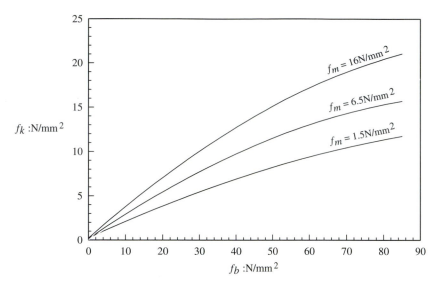

Figure 4.5 Characteristic compressive strength of masonry by Eurocode 6 formula ($K = 0.6$).

Values for the factor K are given in Eurocode 6 ranging between 0.6 and 0.4 according to the type of the masonry. The normalized compressive strength (f_b) is the mean dry strength of the unit multiplied by a shape factor δ to give the equivalent cube strength. Again Eurocode 6 includes a table of values for this factor which are quite closely represented by the formula:

$$\delta = \left(\frac{b}{\sqrt{A}} \right)^{0.37} \tag{4.2}$$

where

δ = shape factor
b = height of the unit
A = loaded area

Figure 4.5 shows for a particular type of masonry a set of curves relating characteristic compressive strength to unit strength for different mortar strengths derived on the basis of the above formulae.

The characteristic strength is a measure of the ultimate strength of the masonry and is divided by a partial safety factor to give the strength to be used in the design of a wall. This factor is essentially intended to provide for possible reductions in the strength of the material in the structure as a whole as compared with the characteristic value. Different values are ascribed to it according to the standards of manufacture of units and of workmanship on site, generally in the range 2.0–3.5.

4.3.2 *Allowance for eccentricity and slenderness*

The cross-sectional area of a wall, whose height was relatively small compared to its thickness, multiplied by the design strength of the masonry would give the total load,

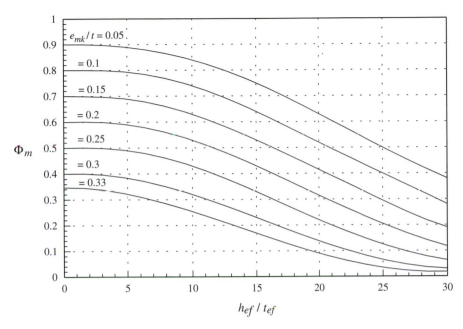

Figure 4.6 Graph showing values of reduction factors (Φ_m) against slenderness ratio for different eccentricities according to Eurocode 6.

centrally applied, which could safely be transmitted by the wall. However, as the height of a wall of given cross-section is increased, its loadbearing capacity is reduced by additional bending stresses. Also, if the load is applied off the central axis of the section lateral deflections and bending stresses rapidly increase leading to further reduction in loadbearing capacity. It is therefore necessary to have available a method whereby allowance can be made for the relative height of the wall and eccentricity of the load. Relevant theories have been produced, some of considerable mathematical complexity, others less rigorous but more easily understood. Whatever the basis, however, for masonry wall design a set of capacity reduction factors is produced which gives the adjustment necessary to allow for any given combination of slenderness and eccentricity. The reduction factors (Φ_m) given in Eurocode 6 are shown in Fig. 4.6.

The parameter defining slenderness is the ratio of the effective height to effective thickness. To understand the meaning of effective height reference may be made to the behaviour of the idealised compression member shown in Fig. 4.7. According to elementary column theory, the relative strengths of columns of the same height (a) hinged at both ends, (b) fixed (i.e. prevented from rotation) at both ends (c) fixed at one end and hinged at the other, (d) fixed at the base and free at the top are in the proportion $1.0 : 4.0 : 2.04 : 0.25$. Strengths in the same proportion would be achieved by hinged end columns of heights in the proportion $1.0 : 0.5 : 0.7 : 2.0$. As shown in Fig. 4.7 these define effective heights and this concept is used in the design of walls to allow for different end conditions. In practice idealised hinged or fixed ends

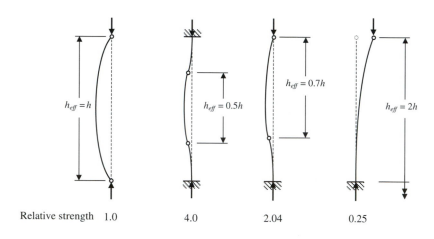

Relative strength   1.0   4.0   2.04   0.25

Figure 4.7 Compression members with different end conditions: deflected forms, relative strengths and effective heights.

do not exist and effective heights are rather arbitrarily defined in codes of practice. For example, the effective height of a wall compressed between concrete floor slabs is conventionally taken as three quarters of the actual height.

In most cases the effective thickness of a wall is the same as the actual thickness but for the commonly occurring cavity wall an empirically defined equivalent thickness is used. In Eurocode 6 this is given as the cube root of the sum of the cubes of the individual leaf thicknesses. The ratio of effective height to effective thickness then gives the slenderness ratio for the wall. In practice, this is limited to about 25 as very slender walls have very low bearing capacity in relation to their area and are only favoured in exceptional circumstances.

In walls of non-rectangular cross-section the effective thickness is not meaningful and it may be necessary to adopt capacity reduction factors which are related to the radius of gyration of the section (i.e. the square root of the moment of inertia divided by the area of the cross section).

It is next necessary to estimate the eccentricity of the load on the wall. As well as additional eccentricity resulting from the lateral deflection of the wall, which develops as the load is applied, eccentricity arises from two other sources. The first is designated 'structural eccentricity' which depends on the position of the resultant loads at the top and base of the wall (Fig. 4.8). In general, the resulting eccentricity varies along the height of the wall and usually has to be found at the mid-height of the wall. It is possible for the mid-height eccentricity so calculated to be very small so that a minimum value of 0.05 of the wall thickness is specified. The second type of eccentricity originates from departures from ideal assumptions of straightness, verticality and uniformity of material. Some design codes assume that allowance for these effects is included in the partial safety factor whilst others, such as Eurocode 6, require the inclusion of an arbitrary 'accidental eccentricity' of the order of height/450. This code also provides for an increase of eccentricity with time (creep effect) for certain types of masonry.

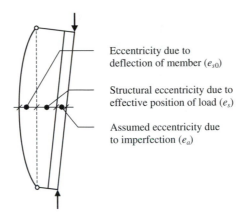

Eccentricity due to
deflection of member (e_{s0})

Structural eccentricity due to
effective position of load (e_s)

Assumed eccentricity due
to imperfection (e_a)

Figure 4.8 Total eccentricity at mid-height of wall ($e_{mk} = e_{s0} + e_s + e_a$).

Having determined the slenderness ratio and load eccentricity, the capacity reduction factor is found from curves such as those in Fig. 4.6.

4.3.3 Loadbearing capacity of walls

The design strength of a wall per unit length under vertical loading can now be found by multiplying the design strength of the masonry by the capacity reduction factor and the thickness of the wall. This is compared with the design load on the same area and if it is greater than this the design is safe.

Alternatively, the calculation can be carried out by equating the design strength to the design load and finding the required characteristic strength:

$$f_k = N_d \gamma_m / \Phi_m t \tag{4.3}$$

where

f_k = characteristic compressive strength of masonry
N_d = design vertical load
γ_m = material partial safety factor
Φ_m = capacity reduction factor
t = thickness of the wall

Knowing f_k it is then possible to select an appropriate unit/mortar combination.

The above is an outline of the method of designing a masonry wall carrying predominantly vertical loading. Essentially the same procedure is used in verifying the adequacy of an existing wall in which case the problem will be to estimate the compressive strength of the material.

4.4 Resistance of walls to in-plane horizontal loading

As described in 4.2 above the wind loads on a building as a whole are resisted by suitably disposed walls acting together. The wind forces will give rise to bending moments and horizontal (shear) forces in the walls, which have to be determined by

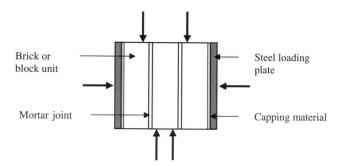

Figure 4.9 Triplet shear strength test arrangement.

an appropriate analytical procedure. The bending stresses are taken into account in designing for vertical loading whilst the horizontal forces are resisted by the shear strength of the masonry.

The resistance of a masonry wall to this combination of vertical and horizontal loads, referred to as 'racking shear', is complex involving the shear and diagonal tensile strength of the material. However, for practical purposes it is usual to calculate the average stress obtained by dividing the design wind load by the plan area of the wall resisting and comparing it with the design shear strength. The latter depends not only on the adhesion between mortar and unit but also on the compressive stress on the joint and is defined by the formula:

$$f_{vk} = f_{vk0} + 0.4\sigma_d \tag{4.4}$$

where

f_{vk} = characteristic shear strength of masonry
f_{vk0} = shear strength under zero compressive stress
σ_d = design compressive stress perpendicular to the shear plane at the level under consideration

The factor 0.4 is a quasi friction coefficient and may vary from one type of masonry to another. Appropriate values for f_{vk0} are given in codes of practice based on accumulated tests on walls and panels or may be determined by tests on small specimens of the type shown in Fig. 4.9. As in the case of compressive strength, the characteristic strength in shear is divided by a partial safety factor to give the design strength.

4.5   Laterally loaded walls and panels

Non-loadbearing external walls and cladding panels are subjected to wind pressure and suction and their resistance depends on the flexural strength of the material and on the support conditions. Thus walls may be supported on two, three or four sides and the support may provide the equivalent of a simple force or a moment restraint as suggested in Fig. 4.10. Calculation of the panel strength may then be undertaken on the basis of a simplified method using the yield line theory, originally developed for

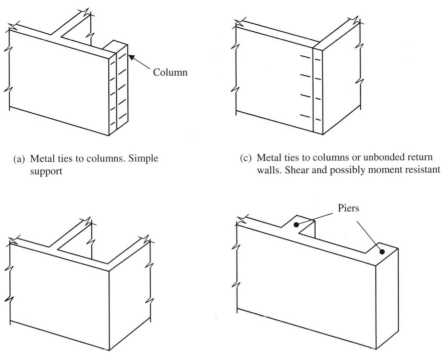

(a) Metal ties to columns. Simple support

(c) Metal ties to columns or unbonded return walls. Shear and possibly moment resistant

(b) Bonded return walls. Restrained support; direct force and moment restraint limited by tensile strength of masonry

(d) Bonded to piers. As (b) for intermediate pier, as (a) for end pier

Figure 4.10 Vertical support conditions for laterally loaded panels. (After BS 5628: Part I.)

reinforced concrete slabs. It is clear that this cannot strictly be applied to a brittle material but good agreement with experimental results has been demonstrated if the ratio of the flexural strengths in the principal directions of the masonry is substituted for the corresponding ratio used in yield line formulae. The flexural strengths of masonry referred to are determined by tests on small panels as in Fig. 4.11. Bending moment coefficients on this basis for a variety of cases are tabulated in BS 5628: Part 1.

An indication of the lateral strength of non-loadbearing walls is given in Fig. 4.12 which shows for various support conditions typical failure pressures against wall area in elevation.

Walls that carry any significant vertical load will have considerably higher resistance to lateral loading than those whose resistance relies on the flexural strength of the masonry. For this reason the lateral strength of loadbearing walls seldom has to be investigated in relation to wind forces but it may be necessary to verify that a particular wall is capable of remaining in place following a gas explosion or other accidental occurrence. Figure 4.13 illustrates the enhancement of the lateral strength of 220 mm and 105 mm thick walls with increase in precompression. The British Code

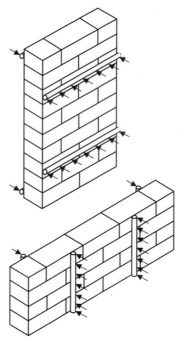

Figure 4.11 Wallettes test specimens for flexural strength of masonry. (After BS 5628: Part I.)

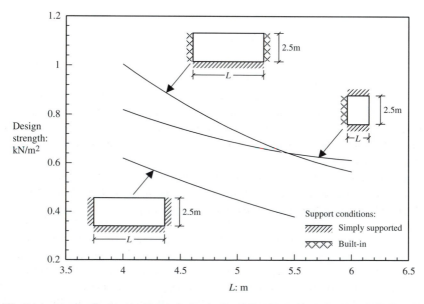

Figure 4.12 Design strength of typical laterally loaded cavity walls according to BS 5628: Part I showing the effect of horizontal span and support conditions. Outer leaf 102 mm brickwork; inner leaf 100 mm blockwork.

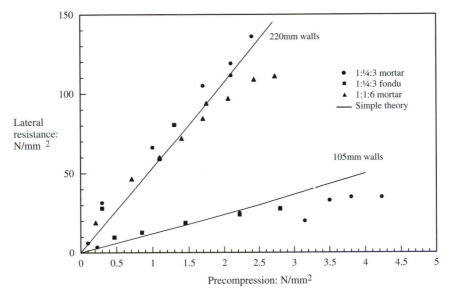

Figure 4.13 Lateral resistance of brick walls with precompression. Collected test results with various mortar strengths compared with simple theory.

requires strength of $34\,kN/m^2$ for a wall to qualify as a 'protected' member in the context of accidental damage.

4.6 Reinforced and prestressed masonry

The compressive strength of masonry is generally adequate for walls in which the load to be transmitted is vertical and compressive stresses predominate. However, the brittle nature of the material does place restrictions on its use where there are relatively large lateral loads resulting in significant tensile stresses. This can arise for example in external walls beyond a certain size, in retaining walls and where seismic forces are anticipated. In such situations it may be possible to use reinforced or prestressed masonry.

There are three basic methods of placing reinforcement in brickwork, as shown in Fig. 4.14, (a) in the bed joints, (b) in specially formed pockets and (c) embedded in concrete between the leaves of a cavity wall.

Similar possibilities exist for blockwork where there is greater scope for the use of special blocks and for the placing of reinforcement in the cores of hollow blocks.

The most effective applications of reinforced brickwork have been for wind resistant panels in the form of light bed-joint reinforcement, for pocket type retaining walls and for grouted cavity walls and beams. Where reinforcement is protected from corrosion by mortar it will, in external walls at least, require to be of austenitic stainless steel. In the other two applications mentioned, and in blockwork, the

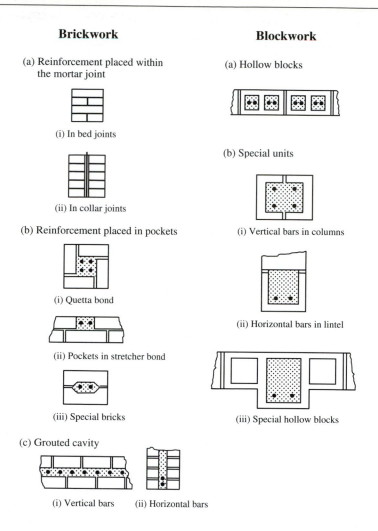

Brickwork

(a) Reinforcement placed within the mortar joint

(i) In bed joints

(ii) In collar joints

(b) Reinforcement placed in pockets

(i) Quetta bond

(ii) Pockets in stretcher bond

(iii) Special bricks

(c) Grouted cavity

(i) Vertical bars   (ii) Horizontal bars

Blockwork

(a) Hollow blocks

(b) Special units

(i) Vertical bars in columns

(ii) Horizontal bars in lintel

(iii) Special hollow blocks

Figure 4.14 Methods of reinforcing brickwork and blockwork.

reinforcement will be embedded in concrete and, provided that the cover is adequate, normal mild or high tensile steel can be used.

The structural behaviour of reinforced masonry is similar to that of reinforced concrete and with appropriate values of the material properties and safety factors the same design methods can be used. Figure 4.15 is a typical diagram that can be used for the flexural design of reinforced masonry elements assuming certain specified values of material strengths and partial safety factors. The parameter M_d/bd^2 is calculated, where M_d is the design moment and b and d the width and effective depth of the beam, and the required steel area ratio for the selected masonry characteristic strength is read off the diagram. It will be noted that cut off lines are shown for compressive and shear failure. The former is undesirable on account of the brittle nature of masonry and therefore excluded. If a steel ratio above the shear cut-off is

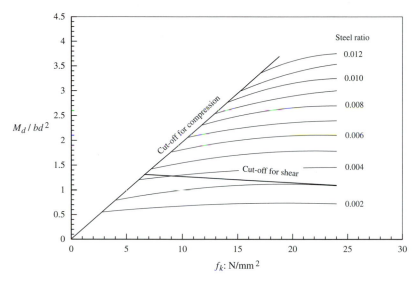

Figure 4.15 Design aid for reinforced masonry beams ($f_y = 460\,\text{N/mm}^2$).

selected it will be necessary to provide shear reinforcement, which is possible in certain types of reinforced masonry including grouted cavity and some forms of hollow blockwork.

It will be apparent that reinforced masonry has the inherent disadvantage that only part of the section will be in compression, thus making limited use of the compressive strength of the masonry. This somewhat inefficient use of the material can be overcome by employing prestressing, again following the same principles as used for concrete. The same sections as in Fig. 4.14 can be used, the prestressing tendons being placed in plastic sheaths or they may be grouted after stressing. Alternatively, high tensile bars can be employed and the use of non-metallic tendons has been explored.

A particular application of post-tensioned masonry which has been developed is for the construction of the walls of sports halls and similar buildings in which a light steelwork roof structure covers a large open floor area. In these buildings, the walls are relatively high and of cellular or T-section (referred to as diaphragm or fin-walls, respectively). Whilst such walls can be built in unreinforced masonry, prestressing greatly extends their possibilities.

In this type of building the roof structure is designed to act as a plate, transmitting wind forces on the sides of the building to the end walls and providing a prop at the top of the sidewalls. The latter are subjected to a combination of lateral and vertical loads and prestressing forces, as indicated in Fig. 4.16. The wall is designed essentially as a propped cantilever in which the total vertical forces are sufficient to prevent the development of tensile stresses in the masonry. It is also necessary to check the shear stresses in the diaphragms and to make allowance for loss of prestress but both design and construction are quite straightforward.

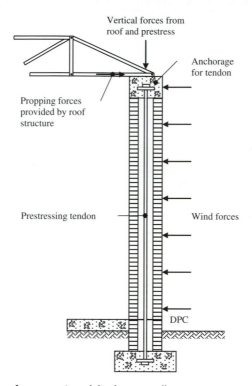

Vertical forces from
roof and prestress

Anchorage
for tendon

Propping forces
provided by roof
structure

Prestressing tendon

Wind forces

DPC

Figure 4.16 Cross section of post-tensioned diaphragm wall.

Further reading

Hendry, A. W. (1998) *Structural Masonry*, 2nd Ed., Macmillan, London.

Hendry, A. W., Sinha, B. P. and Davies, S. R. (1997) *Design of Masonry Structures*, 3rd Ed.,
E. & F. N. Spon, London.

Orton, A. (1992) *Structural Design of Masonry*, Longman, London.

Hendry, A. W. (Ed.) (1991) *Reinforced and Prestressed Masonry*, Longman, London.

Safety of large masonry walls, Building Research Establishment, BRE Digest 281 13/98: Parts 1
and 2, 1998.

Limit state philosophy: partial safety factors and the design of walls for compression and shear,
Brick Development Association, BDA Publ. DGI9, 1988.

Eurocode 1: the code for structural loading, Building Research Establishment, BRE Information
Paper IP 13/98, 1998.

Eurocode for Masonry, ENV 1996-1-1: Guidance and Worked.

Examples, British Masonry Society, BMS Special Publication No. 1, 1997.

Chapter 5

Non-structural aspects of design

5.1 Design for movement

5.1.1 Causes of movement in masonry buildings

Movement in masonry buildings, which may result in cracking, can arise from the following causes:

1. Moisture changes in materials.
2. Temperature changes.
3. The effect of applied loads.
4. Chemical reactions in materials.
5. Inadequate foundation conditions.

Dimensional changes take place in all masonry materials with alteration of moisture content. These may be irreversible following manufacture or may take place with variation of environmental conditions throughout the life of the building. Thus kiln dry clay bricks show an initial irreversible expansion for a short time together with dimensional change following variation in moisture content of the masonry. The extent of these movements in clay brickwork, depending on the type of brick, is within a range of about quarter to half of a millimetre per metre. Dense concrete and calcium silicate products, on the other hand, have an initial shrinkage of a similar order and aerated, autoclaved blocks show somewhat higher movements. Table 5.1 gives illustrative figures for moisture movement of various types of brickwork and blockwork. Information on moisture movement in stone masonry is very limited and would depend on the type of stone and on the relative amount and composition of the mortar. The figure shown in Table 5.1 is thus a very approximate indication for a sandstone masonry with bed joints of thickness about one eighth of the height of the stones.

Thermal movements depend on the coefficient of expansion of the material and the range of temperature experienced by the building element. Indicative values of the coefficient of expansion are given in Table 5.1. The temperature range experienced by a heavy exterior wall in the UK has been quoted as from $-20°C$ to $+65°C$ but there are bound to be wide variations depending on colour, orientation and other factors. In estimating thermal movements, it is also necessary to assume a datum temperature from which these movements take place, e.g. $10°C$ is regarded as a suitable value in the UK.

Table 5.1 Moisture and thermal movement indices for masonry materials

Material	Reversible moisture movement (%)	Irreversible moisture movement (%)	Coefficient of thermal expansion/°C ($\times 10^{-6}$)
Clay brickwork	0.20	+0.02 – 0.07	5 – 8
Calcium silicate brickwork	0.01 – 0.05	−0.01 – 0.04	8 – 14
Concrete brick or blockwork	0.02 – 0.04	−0.02 – 0.06	6 – 12
Aerated autoclaved blockwork	0.02 – 0.03	−0.05 – 0.09	8
Sandstone masonry	0.04	0.005	10

Movements due to loading may result from stressing of the masonry, which may be significant in multi-storey buildings, and may develop either immediately after the application of the loads (elastic deformation) or over a period of time (creep).

Structural movements in adjoining members may affect masonry walls: for example the deflection of supporting beams may induce tensile stresses in the supported wall or horizontal movements in a beam supported by masonry walls may result in cracking in the latter. Again, walls beneath beams or slabs but not intended to support them, may become loaded as a result of the deflection of these elements, resulting in damage to the wall.

Although masonry materials are relatively stable some chemical changes can affect dimensional stability. Thus under certain conditions carbonation of open textured concrete products and mortar can result in additional shrinkage to the extent of about 25% of the free moisture movement. Dense concrete and aircrete units are, however, not significantly affected. Portland cement mortar is subject to attack by dissolved sulphates resulting in disruption of masonry. This however is in the category of a defect rather than a 'movement' which can be allowed for by suitable detailing.

Unsatisfactory foundation conditions are a common cause of cracking in masonry walls. Such conditions requiring particular care include shrinkable clay soils, mining subsidence and filled ground. In certain areas uplift of foundations resulting from particular combinations of soil types and climatic conditions may be experienced. It should be noted that damage due to foundation movement follows uneven settlement or uplift for which masonry walls have limited tolerance.

5.1.2   *Methods of accommodating movement*

If movement is suppressed, very large forces can be set up so that at the design stage provision must be made for moisture, thermal and stress related movements to take place without the occurrence of unacceptable cracking. This is achieved by the selection of suitable materials and by careful detailing rather than by calculation.

The first means of controlling the effect of movement is by limiting the length of walls by the insertion of vertical movement joints filled with compressible sealant. The spacing of such joints depends on the type of masonry but should not generally exceed 15 m in clay brickwork, 9 m in calcium silicate brickwork and 6 m in concrete blockwork. Their width in millimetres should be about 30% more than their spacing in metres. Their location will depend on features of the building such as intersecting walls and openings. The type of mortar used has an important bearing on the ability of masonry to accommodate movement, weaker cement and lime mortars being able to permit considerable deformations over time without cracking. Thus a stone masonry wall in weak lime mortar can be of very great length without showing signs of cracking whereas brickwork built in strong cement mortar will have a very much lower tolerance of movement and the provision of movement joints will be essential. In general, it is advisable to adopt mortar mixes which are not stronger than necessary to meet strength and durability requirements.

Certain details, such as short returns (Fig. 5.1(a)) are particularly vulnerable to damage by moisture and thermal movement and cracking may be avoided by the insertion of a movement joint as in Fig. 5.1(b). Parapet walls are exposed to potentially extreme variations of temperature and moisture conditions and may require to be built on horizontal slip joints so that they can expand or contract without restraint from the supporting structure. Long lintels may also have to have slip joints at their support so that horizontal forces are not transmitted into the masonry with the possibility of causing disfiguring cracks.

Vertical movements in masonry are of the same order as horizontal ones and require special consideration when brickwork or blockwork is used as a cladding to concrete structures. The latter are subject to considerable elastic, creep and shrinkage deformations, which necessitate the use of ties sufficiently flexible or otherwise designed to permit relative movement between the masonry and the concrete structure. There will of course be differential movement between a steel frame and masonry cladding and indeed between the inner and outer leaves of a cavity wall.

It is not usually necessary to quantify vertical movements in masonry buildings unless the outer leaf of a cavity or cladding wall is of greater uninterrupted height than three or four storeys. Estimation of these movements requires knowledge of moisture and thermal movements as in Table 5.1 and also of the elastic modulus for the material and a coefficient which gives the ultimate creep strain as a multiple of the elastic strain. Conventionally, the elastic modulus is taken as 500–1000 times the characteristic strength and the creep deformation 1.5–3.0 times the elastic deformation, depending on the type of masonry. Elastic compression of a wall is not easy to estimate with any great certainty but is likely to be small relative to that arising from moisture and thermal effects but creep deformation may be considerable in high-rise buildings.

Table 5.2 gives an indication of the vertical movement of wall ties at storey levels in the outer leaf of a non-loadbearing cavity wall 20 m in height using the material properties and other data shown in the table.

It will be noted that in calculating the elastic movement it has been assumed that when a wall tie is placed, the elastic compression of the wall below the level of that tie will have taken place. Thus when the ties are placed at the top level of the building they will not move vertically due to this cause, as the wall will already have compressed. On the other hand, creep movements will not develop until some time has elapsed after the completion of the wall and therefore the creep movement at the top level will be the result of this effect over the whole height of the wall.

A similar estimate will be required for the inner leaf of a cavity wall, or for the main structure in the case of a cladding wall, to give the total movement which has to be

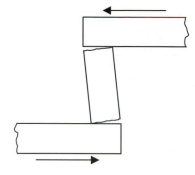

(a) Cracking at a short return due to masonry expansion

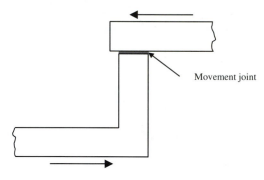

Movement joint

(b) Cracking avoided by insertion of a movement joint

Figure 5.1 Movement at a short return.

Table 5.2 Vertical movement of wall ties (in mm) at storey levels in an eight storey (20 m height) masonry wall

Storey	8	7	6	5	4	3	2	1
Shrinkage	−1.3	−1.1	−1.0	−0.8	−0.6	−0.5	−0.3	0
Relative moisture movement	−4.8	−4.2	−4.0	−3.0	−2.4	−1.8	−1.2	−0.6
Elastic compression	0	−0.3	−0.4	−0.6	−0.8	−0.9	−1.1	−1.2
Creep	−7.4	−6.7	−5.9	−5.2	−4.5	−3.8	−3.0	−2.3
Thermal movement	−7.2	−6.2	−5.3	−4.3	−3.4	−2.4	−1.5	−0.5
Total	−20.7	−18.5	−16.6	−13.9	−11.7	−9.4	−7.1	−4.6

Irreversible shrinkage of masonry = 0.00525%.
Reversible moisture movement from dry to saturated state = 0.04%.
Saturation at time of construction = 50%.
Elastic modulus of masonry = 2100 kN/mm^2.
Creep movement = 1.5 × elastic deformation.
Coefficient of thermal expansion = 10 × 10^{-6}/°C.
Maximum temperature of wall = 50°C.

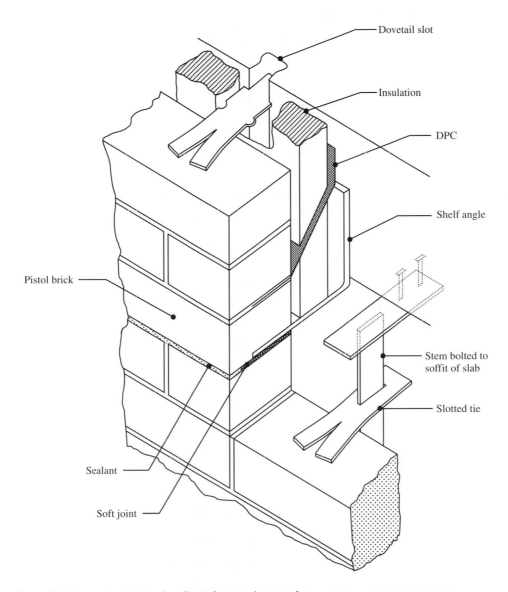

Figure 5.2 Dovetail and slotted wall ties for attachment of masonry to a concrete structure.

allowed for in selecting a suitable type of wall tie. Normal wall ties will be suitable for wall heights up to about 10 m but if this is exceeded, special ties and fixings to provide lateral support such as the dovetail or slotted types shown in Fig. 5.2 will be required.

If masonry panels are built into a reinforced concrete frame compressible joints will have to be provided between the top of a wall and the underside of any intersecting horizontal element, with support at the top of the wall being provided by slotted ties (Fig. 5.2). Infilled panels in steel frames can, however, be built into and tied to the steelwork provided that eccentric loads and short returns are avoided.

5.2   Moisture exclusion

The prevention of moisture penetration is a critical factor in the design of masonry walls, requiring careful selection of materials in relation to exposure conditions, correct detailing and achievement of a good standard of workmanship.

The first step is thus the assessment of exposure conditions with reference to driving rain for which a method, applicable to the UK, is set out in BS 8104: 1992. This is based on maps covering the country showing the quantity of wind driven rain falling on vertical surfaces during the worst expected spell of bad weather. These spell indices are related to six categories of exposure, ranging from 'Very severe' to 'Very sheltered' and in BS 5628: Part 3 recommended thicknesses of single leaf masonry are related to these categories. As an example, most areas of the UK are categorised as 'Moderate' for which a minimum thickness of rendered clay brickwork of 215 mm is suggested. Selection of cavity wall construction appropriate to exposure categories is more complicated and requires consideration of material e.g. clay brick, concrete block etc, mortar designation, joint finish, cavity width and presence or absence of cavity insulation as may be seen from the examples shown in Table 5.3.

Surface characteristics of a wall, whether inherently those of the units or resulting from applied finishes, are also important. Thus certain clay bricks are impermeable and a variety of paints and silicone products are available which aim to reduce surface absorption. In both cases the impermeable surface will increase the liability of water to enter the masonry through imperfectly filled joints. Rendering is effective in protecting a wall provided that it is suitable in terms of composition, thickness and application so as to avoid cracking which would permit the penetration of moisture. Furthermore, impermeable finishes and rendering may result in higher moisture content in the wall by preventing evaporation of such water as may have found its way into it.

Table 5.3 Maximum exposure categories for various types of masonry (after BS 5628: Part 3)

Type of masonry	Mortar type	Designation	Form of Joint	Cavity insulated (Yes/No)	Maximum exposure category
Clay brickwork	C : L : S	(i), (ii) or (iii)	A or B	N	Severe
	C : S + P	(ii) or (iii)	A or B	N	Moderate/severe
	C : S + P	(ii) or (iii)	A or B	Y	Sheltered/moderate
	C : L : S	(ii) or (iii)	C	Y	Sheltered
Dense concrete blockwork	C : L : S	(iii)	A or B	N	Moderate/severe
	C : L : S	(iii)	Rendered	N	Severe
Concrete brickwork	C : L : S	(iii)	A or B	N	Severe
	C : S + P	(ii) or (iii)	A or B	N	Moderate/severe
Calcium silicate brickwork	C : L : S	(iii)	A or B	N	Severe

Mortar type: C : L : S, Cement : lime : sand; C : S + P, Cement : sand with plasticiser.

Mortar designation	Cement : lime : sand	Cement : sand with plasticiser
(i)	$1 : 0 - \frac{1}{4} : 3$	—
(ii)	$1 : \frac{1}{2} : 4 - 4\frac{1}{2}$	1 : 3–4
(iii)	$1 : 1 : 5 - 6$	1 : 5–6

Form of joint (see Fig. 8.1): A, weathered; B, bucket handled; C, raked.

Certain architectural features, such as overhangs and drips, are advantageous in keeping rain water off a wall. On the other hand, large areas of glazing or impermeable cladding can lead to excessive quantities of water running on to the masonry thereby increasing the possibilities of rain penetration and frost damage. Omission of protective features can lead to disfiguring staining of masonry by algal or other organic growth.

Having selected the materials and form of construction of a wall it is then necessary to incorporate damp-proof courses (DPCs) to prevent ground moisture from rising into the wall and the penetration of rain water at openings and at roof level. In cavity walls, it has to be assumed that rain will penetrate the outer leaf. To ensure that water does not find a path bridging the cavity, cavity trays have to be built into the wall wherever this might occur. These are designed to divert water which may have entered the cavity towards the outer leaf. A few DPC and cavity tray systems are shown in Figs 5.3–5.7. These are straightforward examples but in some cases, for example where masonry is supported by a structural frame, arrangements can become complex and difficult to build.

5.3 Durability of masonry structures

5.3.1 General

For practical purposes, durability may be regarded as the ability of a material or construction to remain serviceable for an acceptable period without excessive or unexpected maintenance. What constitutes an acceptable period is not easily defined but the majority of buildings are expected to have a life of many decades. Masonry in its various forms is inherently stable but some care is necessary at the design stage to select materials having properties which are consistent with the expected exposure conditions.

Factors which affect durability include frost action, salt crystallisation, sulphate attack and the action of certain biological agencies. All of these involve the presence of water so that measures designed to exclude or minimise the penetration of water into the masonry, as described in the preceding section, will enhance the durability of the masonry. Similarly, architectural features which have the effect of keeping a wall dry will have a positive influence on durability.

5.3.2 Resistance to frost action

Frost damage is caused by freezing of water in the pores of the material. Thus ice forming first at the surface of a masonry unit entraps water in the sub-surface layers and as this freezes and expands, pressure is built up which may be sufficient to cause spalling of the face of the unit. The mechanism of failure is complicated and depends on a number of factors including the pore structure of the material, the degree of saturation and rate of freezing.

The resistance of masonry to frost damage is very variable and, although tests are available for assessing this property, experience is the most reliable guide for any given location. Clay brickwork is generally resistant but neither strength nor water absorption offer a clear guide so that manufacturer's advice should be sought, particularly where the exposure conditions are likely to lead to persistently wet conditions, for

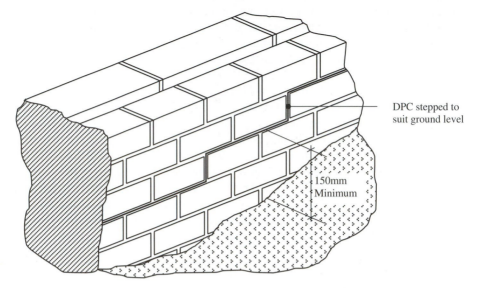

Figure 5.3 Position of DPC above ground level. (After BRE Digest 380.)

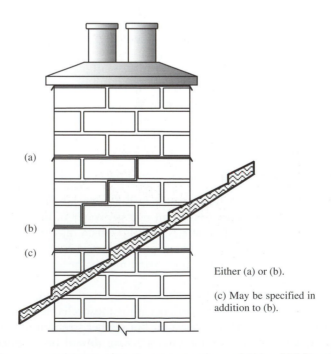

Figure 5.4 DPC in masonry chimney where it penetrates the roof. (After BRE Digest 380.)

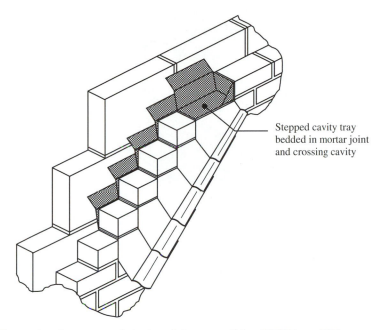

Stepped cavity tray
bedded in mortar joint
and crossing cavity

Figure 5.5 Stepped cavity trays at pitched roof abutment. (After BRE Digest 380.)

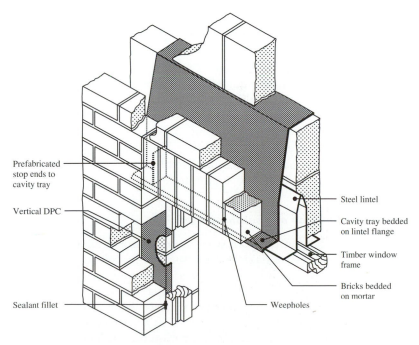

Prefabricated
stop ends to
cavity tray

Vertical DPC

Sealant fillet

Steel lintel

Cavity tray bedded
on lintel flange

Timber window
frame

Bricks bedded
on mortar

Weepholes

Figure 5.6 Cavity tray and DPC systems at a window opening in cavity wall. (After BS 5628:
Part 3.)

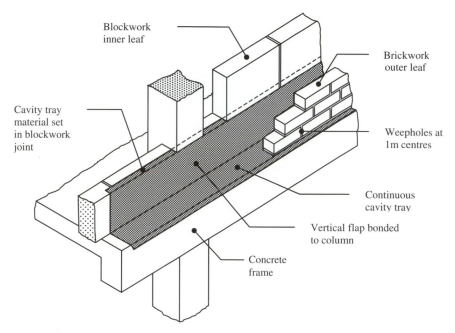

Blockwork
inner leaf

Brickwork
outer leaf

Cavity tray
material set
in blockwork
joint

Weepholes at
1 m centres

Continuous
cavity tray

Vertical flap bonded
to column

Concrete
frame

Figure 5.7 Detail of cavity tray in a concrete frame building with masonry cavity wall cladding. (After BS 5628: Part 3.)

example below ground or in free standing or parapet walls. The mortar mix used for the construction is also important: a $1:1:6$ cement : lime : sand (designation (iii)) is likely to be suitable in an external wall in a sheltered to moderate exposure situation but a stronger mortar would be required where there is a high risk of saturation. Calcium silicate brickwork in similar conditions can be built in $1:1:6$ or $1:2:9$ mortar but this type of masonry is susceptible to damage if exposed to sea-water or salt contaminated water from road de-icing. The resistance of concrete bricks is related to strength and for exposed situations a unit strength of $15 \, \text{N/mm}^2$ is advisable, set in $1:1:6$ mortar. All types of concrete blockwork show good resistance to frost damage and can be built in $1:1:6$ mortar.

5.3.3   Salt crystallisation and sulphate attack

The crystallisation of salts in the pore structure of masonry units can result in significant damage by progressive flaking of the surface. Dissolved salts can be carried into the masonry from ground water, atmospheric pollutants or may originate within the units or in adjacent materials. In warm weather the water evaporates and the dissolved salts crystallise in the pores of the material below the surface to form a hard skin which may then flake off exposing a new surface to the same process. The masonry has to be wet for a considerable period for a salt solution to be formed and for the evaporative mechanism to be established. Damage from this cause is more common in warmer climates and in magnesian limestone.

Damage from salt crystallisation is essentially a physical process whereas sulphate attack results from chemical reactions between cement mortar and dissolved sulphates, which may originate either in the ground water or in clay bricks. The sulphates react with certain constituents of the mortar resulting in the formation of calcium sulpho-aluminate which occupies an increased volume thus resulting in the disruption of the mortar and failure of the brickwork by splitting along the joints.

Liability to sulphate attack is increased if the wall remains wet for a prolonged period and is minimised by correct selection of materials and protection from ingress of water. Brickwork below damp-proof course level may unavoidably be wet and in this case particular attention has to be given to the selection of the bricks. If the ground water contains sulphates, a sulphate resistant cement will be necessary. The outer leaf of a cavity wall is also vulnerable in exposed situations, especially if the cavity is insulated, as this will slow down the evaporation of moisture following a spell of wet weather. If such a wall is protected by rendering, rain penetration will be reduced but if, as a result of some defect, water does enter the masonry drying will be retarded and the possibility of sulphate attack increased. Furthermore, expansion of the mortar beds will lead to cracking of the rendering and further penetration of water.

5.3.4   Atmospheric pollution

The main source of atmospheric pollution is the burning of fossil fuels which results in the presence of sulphur and nitrogen acids. Sulphur dioxide is a very widespread and damaging pollutant as it combines with water to produce sulphurous acid which attacks tricalcium aluminate in cement with the effects described above. Certain types of natural stone which have a pore structure comprising many pores of small diameter (micropores) are particularly susceptible to damage by pollution. These pores retain acidic moisture from the atmosphere resulting in the surface layers of the stone being subjected to prolonged attack. This in turn leads to the formation of salt crystals and consequent damage similar to that caused by crystallisation of salts from internally transmitted solutions as described in the preceding paragraph. The degree of micro-porosity of a stone is thus a predominant factor in relation to its durability.

5.3.5   Attack by biological agencies

A large variety of algae, lichens, mosses, fungi and even bacteria as well as higher plants can establish themselves on the surface of a masonry wall. These organisms can penetrate the pores of masonry materials and may cause damage by generating organic acids with similar effects to atmospheric pollution. The presence of organic growth on the surface of masonry therefore has a negative effect on durability. All such growths, however, require the presence of water for their development and the first line of defence against deterioration from this cause is to keep the wall dry as far as possible.

5.3.6   Durability of metal components embedded in masonry

Although masonry is comparatively highly durable it is frequently necessary to incorporate within it metal components such as wall ties, fixings, lintels and reinforcing bars. As metals are chemically more reactive than masonry materials it is necessary to ensure that such components are capable of withstanding the conditions to which they are exposed without significant deterioration over the expected life of the masonry. This is achieved by correct selection of the metal and by such protection from corrosion as may be practicable.

The most common metal components are wall ties in cavity wall construction and these are often exposed to very severe conditions being partly embedded in a thin outer leaf of masonry, frequently saturated for prolonged periods. For sheltered situations galvanised low carbon steel ties will be suitable but for severe conditions austenitic stainless steel or non-ferrous metal will be required. Galvanised low carbon steel will generally be adequate for internal fixings which are unlikely to be exposed to wet conditions. It should be noted that components such as twisted wire wall ties are of rather thin material and are often exposed to very severe conditions which means that any failure of the galvanising can quickly lead to complete failure of these components with serious implications for the stability of the wall. Such wall ties must therefore unquestionably conform with the relevant standard and, if any doubt exists as to their adequacy, 18/8 austenitic stainless steel ties should be adopted—the increased cost will be marginal in the cost of the wall.

Light gauge bed joint reinforcement is sometimes used with a non- or semi-structural function in masonry walls. The first application is aimed at crack control particularly around openings where cracks may result for example from shrinkage, settlement or other movements. Structural use may be for strengthening walls against lateral loads or for lintels. In both cases the material used has to be adequate for the exposure conditions and follows the same general lines as for wall ties. It must be noted that mortar will not provide long term protection of embedded low carbon steel. This depends on the existence of an alkaline environment around the metal which will exist for some time after construction but which will become acid as a result of carbonation of the mortar. Exposed walls with bed joint reinforcement will therefore require the selection of stainless steel.

In fully reinforced structural masonry, similar considerations apply but in this case the steel will in many cases be embedded in concrete which permits the use of low carbon steel provided that adequate cover is provided for the exposure conditions. The cover recommended in BS 5628: Part 2 varies from 20 mm to 60 mm according to the exposure and the concrete grade. If austenitic stainless steel, or carbon steel with a minimum 1 mm protective layer of stainless steel, is used, no minimum cover is specified although some cover will be required for the development of bond.

5.4   Thermal design

5.4.1   General principles

The objective in the thermal design of a building envelope in temperate and cold climates is to achieve an acceptable standard of comfort internally with as small an expenditure of energy as practicable. Maximum possible energy efficiency, however, will not usually be a practical aim as it could require very high construction costs and could conflict with other functional requirements of the building. As public interest in energy conservation has developed for economic reasons and as a means of limiting the increase in 'greenhouse' gases, increasingly strict criteria have been laid down in

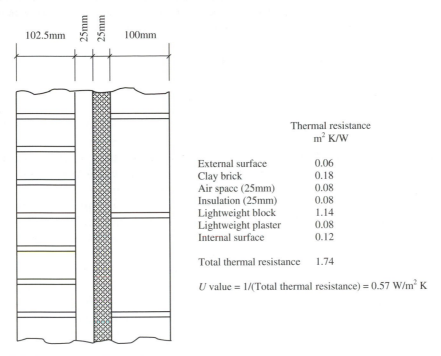

102.5mm   25mm   25mm   100mm

Thermal resistance
m² K/W

External surface	0.06
Clay brick	0.18
Air spacc (25mm)	0.08
Insulation (25mm)	0.08
Lightweight block	1.14
Lightweight plaster	0.08
Internal surface	0.12
Total thermal resistance	1.74

U value = 1/(Total thermal resistance) = 0.57 W/m² K

Figure 5.8 Calculation of U value for a cavity wall.

building regulations for the thermal performance of buildings. These, rather than theoretically possible levels of performance, are used in design.

Since energy efficiency depends on, amongst other factors, the ability of the whole envelope to retain internal heat, it is obvious that assessment of heat loss must involve consideration of all the elements which comprise it of which the external walls are an important part. The requirements set out in building regulations may be met by ensuring that specified values for the thermal transmittance of individual elements are achieved or by adopting a 'target' value for this quantity for the envelope as a whole. Thermal transmittances, known as U values, are defined in terms of watts/ square meter/degree Kelvin (W/m² K) and can be calculated from standard material properties as in the example shown in Fig. 5.8.

It will be seen from this example that the thermal resistance for the wall construction is first obtained, including an allowance for internal and external surface resistance. The resistances for the various parallel components are added together and the thermal transmittance is given by the reciprocal of this figure. Numerous variations of cavity wall construction are available depending on the type and location of the insulation and on the thickness and materials used for the leaves. The same principles can of course be applied to solid walls.

The calculation as shown applies to a cavity wall in which the leaves are completely separated but in fact they have to be interconnected by, usually, metal ties and the cavity may be bridged by other components which have the effect of reducing the thermal resistance of an area of wall. This may also be affected by air movement in the cavity and by the moisture content of the masonry. Adjustments to the effective U value for the construction will therefore be required to allow for these effects.

Where a target U value is adopted for the building variation in the element values is possible. That is to say, somewhat higher values than the target figure can be accepted in certain areas, compensated by lower values elsewhere.

5.4.2 Thermal insulation

Achievement of currently acceptable U values for masonry wall construction is difficult without additional insulation other than by the use of rather thick blocks with particularly good thermal properties, which in fact is standard practice in many countries. If thin wall masonry is used, the additional insulation may be internal, external or within the cavity of a cavity wall and many different materials are available for use in these situations. The location of the insulation is important in relation to thermal behaviour of the wall. Thus if the insulation is internal, the wall will be relatively cold and will thus retain on average a higher moisture content with reduced thermal resistance. External insulation, on the other hand, will result in the masonry being at a higher average temperature and drier. Its insulating properties will thus be improved but heat will be absorbed from the interior to attain this higher temperature. The effect of 'thermal mass' plays a part here since heat stored in the wall will tend to reduce fluctuation of heat within the building whereas relatively light internal insulating material will result in more rapid response to temperature change within the building. Which of these characteristics is the more advantageous will depend on the use of the building—constant occupation will benefit from the stabilising effect of the thermal mass of the walls whereas with intermittent occupation, internal insulation with quick response to heating and a small amount of heat absorbed by the fabric will be more economical in energy use.

In a more general way, thermal mass has an important influence on the response of a building to variation of ambient conditions which include solar radiation as well as air temperature and movement. This is a dynamic rather than a steady state situation as usually assumed in the calculation of heat loss and insulation requirement. Representation of dynamic thermal behaviour is possible and in certain circumstances could lead to economies in fuel use but is of course considerably more complex and therefore not usually attempted.

Other factors influencing the selection of an insulating system are those of space and appearance. Internal insulation may require as much as 50 mm of sheet material secured to timber battens on the wall, thus reducing the useable internal space. External insulation will not have this disadvantage but will influence the appearance of the building and is likely to require protection by sheet material, weatherboard, tiling or rendering.

Cavity walling has the advantage that the major insulating layer can be contained within the thickness of the wall, augmenting the value of the inner leaf and plasterboard finish with good insulating properties.

In recent years, special energy efficient designs for houses and other buildings have been developed which make use of combinations of glazing, insulation and the energy storage capacity of masonry walls to achieve a situation where the heating requirements of the building are met by solar radiation with very much reduced need for other means of heating.

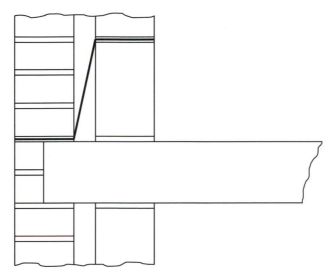

Figure 5.9 Cold bridge formed by slab crossing cavity.

The most commonly used insulation materials include extruded polystyrene, rigid polyurethane and mineral fibre in sheet or roll form. Special fixings are available for securing these materials to the wall face or within the cavity. For the latter situation the insulation can be introduced after construction in the form of beads, fibre or foam. These materials are generally water resistant but if wet their insulating value will be seriously reduced. Joints between sheets of cavity insulation must be taped and gaps in foam insulation must be avoided so that water will not find its way across the cavity.

5.4.3 Condensation

Water will condense on a surface if the temperature of the air in contact with it falls below that at which moisture can remain within it as a vapour. This temperature is known as the 'dew point' and depends on the air temperature and relative humidity. Under winter conditions moisture laden air within a building will move outwards through a wall and if this is uninsulated, or insufficiently insulated, will lead to condensation on the inner surface if this is below the dew point. Such condensation may cause damage to decorations or to mould growth so that it is a primary objective of thermal design to prevent it taking place.

As a precaution against condensation, it is normal practice to incorporate a vapour control layer in the form of a plastic sheet on the warm side of the insulation. There should be as few holes as possible in this sheet with joints sealed with tape in order to avoid local areas of condensation.

Where some material of high thermal transmittance is for some reason carried through the insulating layer or layers into a cold zone, a thermal bridge will be created, reducing the effectiveness of the wall insulation and possibly creating conditions for condensation or pattern staining in the inside of the wall. A classic example is where a concrete floor slab is carried through both leaves of a cavity wall as in Fig. 5.9.

Such a detail was at one time common practice for structural reasons but is now avoided because of the thermal disadvantages. Cold bridges can also be created by the presence of metal lintels and fixings and even by mortar beds in lightweight blockwork. Attention has also to be given to insulating the edges of concrete ground floor slabs where they adjoin external walls.

5.5   Acoustic properties

The ability of a wall to reduce direct sound transmission between two spaces depends primarily on mass and therefore masonry walls are inherently effective in meeting this requirement. However, the question of sound insulation is complicated since sound can be transmitted other than directly through a wall by flanking paths as indicated in Fig. 5.10. Such paths may follow interconnected walls or through air spaces and sound can travel considerable distances horizontally and vertically in this way.

The most common situations in which the sound insulating properties of masonry walls are critical are for walls separating occupancies (party walls) and for external walls exposed to traffic noise. Normal wall thicknesses will generally be adequate for sound insulation in both cases but care must be taken in the construction of party walls to ensure that all joints are properly filled and, if a cavity wall is used, that only light wall ties are employed. If this is not possible for structural reasons a single thickness wall may be preferable. Diminution of the thickness of a party wall, for example by chases, should be avoided and floor joists carried by a party wall should be supported by metal hangers and not built into the wall.

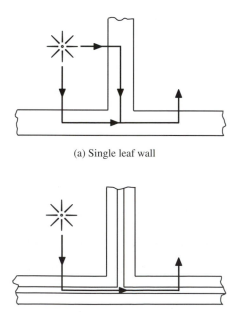

(a) Single leaf wall

(b) Cavity wall: light weight internal leaf, heavy outer leaf

Figure 5.10 Flanking transmission.

To avoid flanking transmission through air spaces in external walls, window and doorframe openings should be sealed to prevent sound penetration.

Although masonry walls are relatively efficient in providing sound insulation, they will reflect sound and thus in some spaces, such as assembly halls, surface treatment with sound absorbing material may be required to provide an acceptable acoustic ambience. At the same time, it should be noted that sound absorbent finishes are not in general effective insulants and cannot be relied upon to provide sound insulation between spaces.

5.6 Fire resistance

Masonry materials are incombustible and therefore inherently effective in providing fire protection for the periods of time specified in building regulations. The objectives of fire protection include provision for escape of occupants and protection of fire fighters, containment of a fire within the building and prevention of its spread to adjoining buildings.

Minimum thicknesses of various types of walling for periods of resistance up to six hours have been determined by standard test procedures and are set out in codes of practice such as BS 5628: Part 3. In the selection of thickness, certain structural requirements apply in order that in the event of a fire the stability of a wall should be maintained for a reasonable period.

In securing the objectives of fire protection, detail design and satisfactory workmanship are essential. Of particular importance is the provision and correct installation of fire stops in cavities and where services pass through a wall. Clearly, the joints in a protective wall must be properly filled and the perimeter details must be such that fire will not simply by-pass the wall.

Fire protection regulations are complex and have to be taken into account at all stages of the design of a building and indeed when work is being carried out to upgrade an existing building.

Further reading

Thomas, K. (1996) *Masonry Walls*, Butterworth–Heinemann, Oxford.

Estimation of thermal and moisture movements and stresses, Building Research Establishment, BRE Digests 227, 228 and 229, 1979.

Climate and site development: influence of microclimate, Building Research Establishment, BRE Digest 350: Part 2, 1990.

Resisting rain penetration with facing brickwork, Brick Development Association, BDA Publ. DN 16, 1997.

Design and appearance, Building Research Establishment, BRE Digests 45 and 46, 1964.

Decay and conservation of stone masonry, Building Research Establishment, BRE Digest 177, 1975.

Control of lichens, moulds and similar growths, Building Research Establishment, BRE Digest 370, 1992.

Brickwork durability, Harding, J. R. and Smith R. A., Brick Development Association, BDA Design Note 7, 1986.

Cavity insulation, Building Research Establishment, BRE Digest 236, 1980.

Surface condensation and mould growth in traditionally built dwellings, Building Research Establishment, BRE Digest 297, 1985.

Interstitial condensation and fabric degradation, Building Research Establishment, BRE Digest 369, 1982.

Improved standards of insulation in cavity walls with an outer leaf in facing brickwork, Ford and Durose, Brick Development Association, BDA, Publ. DN 11, 1982.

Sound insulation of separating walls and floors, Building Research Establishment, BRE Digest 333, 1988.

Sound insulation: basic principles, Building Research Establishment, BRE Digest 337, 1988.

Insulation against external noise, Building Research Establishment, BRE Digest 338, 1988.

Chapter 6

Masonry wall construction

6.1 Types of masonry walls

6.1.1 General

There are numerous types of masonry wall construction which may be classed under the headings of loadbearing or non-loadbearing, internal or external and according to the type and material of the units. A further distinction arises as between walls which are of the same thickness as the units and those which are of greater thickness than this. As described in previous chapters, the relevant design criteria depend on which, or which combination, of these categories a particular wall belongs.

6.1.2 Single leaf walls

The single leaf wall is used internally and externally and, as the name implies, is of the same thickness as the units from which it is built. Blockwork walls are generally of this type although obviously can be widely different in thickness. Single leaf brickwork on the other hand will normally be about 100 mm thick and if external will usually be part of a cavity wall or cladding to a concrete wall or timber frame panel. This thickness of brickwork or blockwork cannot be regarded as proof against rain penetration or to provide in itself sufficient thermal insulation and these functions have to be provided by other components or elements. On the other hand, single leaf blockwork walls of greater thickness can in certain circumstances meet these requirements but may require to be rendered on the outside.

Single leaf walls are laid in what is described as running or stretcher bond in which the units in successive courses overlap by half or quarter of their length as in Fig. 6.1(a). This produces a slightly monotonous appearance unless the surface is broken up by building in panels to give relief, in patterns of coloured bricks or by the use of special bricks to simulate bonds used in thicker walls. Sometimes these walls are laid in 'stack' bond as in Fig. 6.1(b) for architectural effect but it has to be noted that this variation has an inherent structural weakness.

6.1.3 Cavity walls

Cavity walls in UK practice consist of two single leaves, the inner leaf usually being of blockwork, mainly to provide thermal insulation. The leaves are tied together with wall ties 450 mm apart to ensure lateral stability. Damp-proof courses and cavity trays

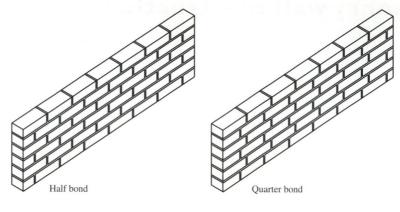

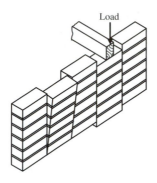

(a) Stretcher bond

Half bond

Quarter bond

Load

(b) Stack bond, showing structural weakness of carrying beam loads

Figure 6.1 Single leaf walls.

have to be incorporated to exclude moisture penetration and in some cases this results in rather complex details.

6.1.4   Bonded brickwork walls

Walls having a thickness equal to or greater than the length of a unit can be laid in a variety of bonds. If the long face of a brick is exposed this is termed a 'stretcher' and if the end face is exposed this is a 'header': the bond pattern results from the arrangement of headers and stretchers visible in the face of the wall. A number of commonly used bonds are shown in Fig. 6.2. Bond patterns are adopted for reasons of appearance and have little effect on the strength of the wall, except in the case of walls built as two thicknesses of stretcher bond which are less strong than bonded walls of the

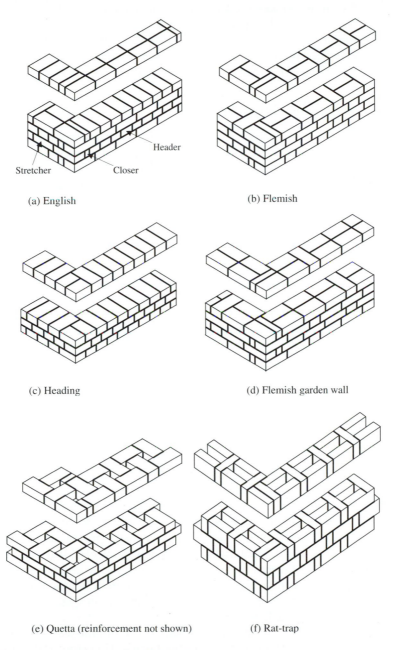

(a) English

(b) Flemish

Header

Stretcher

Closer

(c) Heading

(d) Flemish garden wall

(e) Quetta (reinforcement not shown)

(f) Rat-trap

Figure 6.2 Brick masonry bonds. (After BS 5628: Part 3.)

same thickness unless wall ties are incorporated. This type of wall is sometimes used to produce brickwork which is 'fair faced' on both sides—a result which cannot be obtained with other bonds because the length of a standard brick (header) is slightly less than the combined thickness of two bricks (stretchers) plus a mortar joint.

Bonded walls are no longer used externally in domestic buildings for reasons of thermal efficiency and cost but may be adopted for internal, party walls to provide sound insulation or fire protection. They are, however, often used in larger buildings where they are justified by structural requirements and for architectural effect. Bonded brickwork is also appropriate for free standing and retaining walls.

6.1.5   Special types of walls

Masonry construction is by no means restricted to straight lengths of wall and is frequently used to build curved, circular or indeed any plan shape of wall. Walls of cellular or T-section, either in brickwork or blockwork, known as diaphragm or fin walls have come into use in recent years (although the concept is hardly new as it can be found in ancient stone buildings). These walls are used to support a relatively light roof structure in large open span buildings such as sports halls. Because of their cross-section they can be built to a height of 10 m or more making efficient use of the material, particularly if post-tensioned.

Reinforcement and prestressing can be used to overcome the limitations of masonry in regard to tensile strength. As well as the application to cellular walls mentioned above a successful use of reinforced masonry is in the construction of 'pocket type' retaining walls. In this form of construction vertical cavities extending over the height of the wall are left at intervals in the course of construction. Reinforcement is placed in these cavities and concreted in, only requiring a board to close the cavity until the concrete sets.

6.1.6   Stone masonry walls

Although natural stone masonry is not now used to any extent in new building, it exists in vast quantities and restoration and repair work in this material is therefore in considerable demand.

Stone masonry construction takes many forms, some of which are illustrated in Fig. 6.3. The type of walling depends first on the stones used i.e. whether roughly or accurately shaped. Roughly shaped stones are used for varieties of rubble masonry whilst very accurately cut blocks (sometimes referred to as 'dimension stone') are required for ashlar. The shaping of stone is an expensive process and therefore even moderately accurately worked material is placed only in exposed faces of a wall. Faces which are concealed by finishes or otherwise not seen in the completed building are generally built in rough rubble masonry with a core up to 150 mm in thickness filled with small fragments of stone set in mortar as shown in Fig. 6.4. In this way, incidentally, there is very little waste of material.

Natural stone in the form of thin slabs is used in current practice as the outer leaf of a cavity wall the inner leaf being concrete or blockwork. Thin stone slabs are also used as a veneer to give the appearance of masonry although this does not strictly speaking qualify as masonry as the slabs may be fixed to the backing without mortar between them.

Site work in stone masonry construction is somewhat specialised, requiring rather greater skill on the part of the mason and therefore more expensive than brickwork or blockwork.

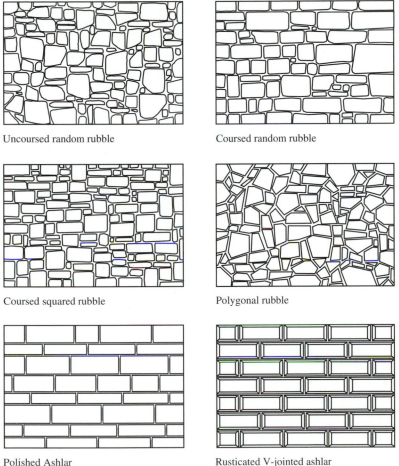

Uncoursed random rubble

Coursed random rubble

Coursed squared rubble

Polygonal rubble

Polished Ashlar

Rusticated V-jointed ashlar

Figure 6.3 Types of stone masonry walls.

6.2 Masonry construction

6.2.1 General

Masonry construction has a history extending over thousands of years and in many ways the basic procedures have changed little. They include procurement and storage of materials, setting out of the building, access to the work, laying of bricks, blocks or stones and finishing of the masonry. In many, but by no means all, old buildings, very high levels of skills were required and achieved in the fashioning of stone masonry units and in the actual construction whatever the material. Work of the highest quality can still be obtained but it is expensive and requires a high level of supervision. However, the majority of masonry wall construction does not now incorporate elaborate architectural features and with simpler detailing the achievement of satisfactory standards is entirely practicable with normally available workmanship and supervision.

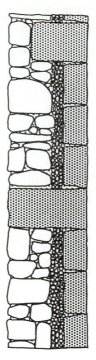

Figure 6.4 Cross section of a stone masonry wall with ashlar face and rubble backing.

However, for masonry to maintain its place as a primary construction material it is not sufficient to rely on past use and performance. All concerned on site must be aware of best current practice and constantly look for improvements in materials and methods of construction. Materials have in fact changed considerably over the past few decades in terms of variety, quality and properties. Design has become more sophisticated and although methods of construction have changed rather more slowly, new techniques are emerging, as will be described later in this chapter.

6.2.2 Current site procedures

Almost all masonry is now built in bricks and blocks delivered to site in polythene wrapped packs of 500 bricks or equivalent number of blocks. This greatly facilitates handling and storage as the packs can be moved from the delivery vehicle by forklift truck or crane close to where they are to be laid (Fig. 6.5). In this way also, units are protected from damage and from saturation by rain. It is, however, necessary to ensure that the floor or scaffolding on which these packs are stored is not overloaded and relevant notices should be displayed on site (Fig. 6.6).

Cement, lime and sand for mortar must be suitably stored, protected against rain and damp conditions. On large sites, mortar may be delivered in premixed form in metal containers or silos (Fig. 6.7), ensuring accuracy of gauging and reducing site

Figure 6.5 Delivery of pack of blocks by crane close to laying area. (Courtesy of Morrison Construction Group plc.)

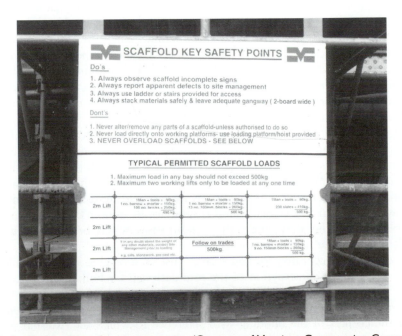

Figure 6.6 Scaffold loading safety notice on site. (Courtesy of Morrison Construction Group plc.)

Figure 6.7 Pre-mixed mortar silos. (Courtesy of Morrison Construction Group plc.)

work. Other components such as wall ties, cavity trays and damp-proof course material must also be securely stored against damage and waste.

If mortar is not supplied in pre-mixed form, it must be accurately gauged, normally employing boxes, and mixed by machine: gauging and mixing by shovel is unsatisfactory except for the most trivial of jobs. After mixing, cement mortars should be used within two hours unless of a specially retarded, ready to use type. Within this period, workability may be restored with the addition of a small amount of water (known as re-tempering) but thereafter should be discarded. Re-tempering may be necessary in warm weather but as far as possible avoided by mixing only as much mortar as can be used within the two hours.

Special precautions are necessary in winter conditions to protect units from becoming excessively wet and particularly to avoid using mortar at temperatures approaching freezing when the setting time is much longer than normal. If the mortar freezes, its structure is disrupted by expansion of the contained water and although it may harden when the temperature rises it is likely to be weaker and more porous than normal. Mortars with less than 1 : 6 cement : sand should not be used in cold weather and the use of a plasticiser rather than lime in these circumstances may be advisable insofar as there will be less water in the mix to freeze.

Setting out points accurately located with the use of surveying instruments and steel tapes will precede the construction of the foundations of a building. Vertical datum points or temporary benchmarks will also have been established prior to the commencement of operations. Profiles, i.e. horizontal boards supported by pegs, will be set up at the corners of the building with reference to which the external walls can be built, intermediate profiles being positioned from which to locate openings and internal walls. Particular care is necessary in building the first courses of masonry to ensure accuracy and all subsequent courses must be laid level and in such a way that the resulting wall is vertical. The designer will have prescribed the required bond pattern and should have dimensioned the building in such a way that the cutting of bricks or blocks is avoided or at least minimised. Vertical joints should be such that there is an overlap of at least a quarter of the length of the unit above and below and the mason must ensure that there is no visible drift of the vertical joints in the height of the wall. It is also important to ensure uniformity of joint thickness by means of a gauge rod and to keep accurate correspondence of the level of bed joints in the leaves of a cavity wall. This is essential in order that wall ties and other fixings should be correctly located: errors in this frequently lead to unacceptable bending of ties to fit them into incorrectly coursed walls. Wall ties should be placed in the mortar bed as it is laid, not pushed in afterwards, and great care taken to avoid excess mortar falling into the cavity. This can lead to expensive trouble in service if water thereby finds its way into the interior of the building and can be avoided by suspending a wooden board in the cavity below the work area.

Brick and block laying requires considerable skill to ensure that all the above requirements are met: that joints are uniform and completely filled and that excess mortar is struck off without smearing the surface of facing brickwork or allowed to fall between the leaves of a cavity wall. Figure 6.8 shows an example of unsatisfactory workmanship in the latter respect. Figure 6.9 is an example of a carelessly built clay block infill wall, showing badly filled and irregular joints.

In brick or block laying, adhesion between the units and mortar has to be ensured, having regard to their characteristics. Some units will absorb a considerable amount of water from the mortar making it difficult to place them accurately and in warm weather may lead to poor adhesion. At the other extreme, too high a water content in the mortar may lead to movement after placing and extrusion from the joints, leading to staining of the face of the work.

In building with larger and heavier blocks special arrangements for lifting and handling may be required. Thus a staging will be required to enable blocks to be set without requiring too great a lift from where they are placed awaiting use (Figs 6.10 and 6.11).

The joints in facing brickwork or blockwork should preferably be finished as the work proceeds rather than being pointed later. If, however, pointing is required for visual effect the joints should be raked out to a depth of 10 to 15 mm as the work proceeds and kept free of dust and loose material to ensure a satisfactory bond. Problems of appearance can arise in facing brickwork as the result of variation in the colour of the bricks, whether intended or not, even within a single delivery. Where such variation is likely to arise, it is common practice to construct a trial panel before the work commences and to keep this in existence on site for reference, the several parties concerned having agreed as to its appearance. It may be necessary to mix the bricks before use to preserve overall uniformity of appearance.

One of the more difficult operations in masonry construction is the correct installation of damp-proof courses, cavity trays and flashings. These details should be clearly thought out by the designer and subject to inspection as failures after completion of the building will give rise to damage to interior finishes and invariably expensive repairs.

Figure 6.8 Accumulation of mortar droppings at a cavity tray.

Figure 6.9 A poorly built blockwork infill wall.

Figure 6.10 Reaching the limit of height and weight for block laying without the use of a staging. (Courtesy Morrison Construction Group plc.)

Figure 6.11 Block laying from a staging. (Courtesy Morrison Construction Group plc.)

As described in Section 6.1.6, stone masonry walls can be built in many different ways, ashlar being equivalent to blockwork where the stones are accurately cut to size off-site and laid in courses. This type of masonry can be used as the outer leaf of a cavity wall, the inner leaf being of concrete blockwork or brickwork. The techniques of construction are similar to those used for cavity walls in these materials. In traditional construction, accurately dimensioned stones laid to courses form the outer face of ashlar walling backed by rubble masonry. The space between the two component walls is filled with small stones set in mortar but a sufficient number of through stones at openings must be provided to bond them together. The joints in the ashlar face are as thin as 5 mm and in building the wall it is usual to use metal spacers to ensure that the required thickness of joint is obtained. The stones in ashlar masonry will often be too heavy to be placed by hand, even with a two handed lift, and therefore suitable lifting tackle or the use of a crane will be required.

Many varieties of rubble masonry can be built ranging from that using squared stones laid to courses with relatively thin joints, to that using random sized and shaped stones with thick joints. The satisfactory construction of random rubble masonry is dependent on the skill of the mason in selecting and laying the stones, having regard to their bedding plane. The wall has to be adequately tied through its thickness by incorporating bonding stones, extending through about two thirds of the thickness, approximately one per square metre of surface area.

Lime mortar is traditionally used for natural stone masonry and if cement mortar is used it must not be too strong in relation to the stone as this may result in severe damage to the latter, by freezing of water trapped in the stones. As in all masonry construction, new work has to be protected by covering with polythene or sacking during periods of cold weather.

6.3 Developments in masonry construction

As already commented, it is essential if masonry is to retain its position as a major construction material in competition with alternatives, for significant improvements in site practice to take place. Long standing criticisms of conventional methods have included that masonry buildings take too long to construct, that it is difficult to find skilled masons and that this is, at least partly, because of unattractive working conditions on site. Efforts to overcome these criticisms have led to experiments with prefabrication, the use of new types of units, improved site methods, organisation and working conditions.

6.3.1 Prefabrication

Prefabrication of brickwork panels was attempted in the UK as early as the 1930s and developed on a commercial scale in the 1960s, mainly in European countries, but with some activity in North America, Australia and South Africa. Various methods for the production of panels have been developed. In some, the units are laid on a soft bed in a horizontal mould and then grouted up following which a layer of insulation is placed over the brickwork and over this an inner concrete leaf is poured. An alternative procedure is to set the units in a vertical mould then fill the joints with mortar under pressure. Automated brick laying in factory conditions has also been attempted.

All these methods require rather expensive equipment but can be undertaken with a limited input of skilled labour. A compromise is to produce prefabricated elements by conventional bricklaying using jigs either in a factory or on site.

One of the major problems in prefabricated construction is that of jointing between the panels, most systems have used synthetic rubber mouldings for this purpose, similar to those used in large panel concrete structures.

The advantages of prefabrication include the following:

1. Work carried out under factory conditions permitting close control of materials and workmanship.
2. High standard of work achieved with available skills.
3. Accelerated site construction.
4. Avoidance of delays due to weather.
5. Reduced requirement for storage of materials on site and reduced waste.

The disadvantages include:

1. High cost of plant and factory space.
2. Long, continuous production runs required for economic viability.
3. Special transportation and heavy lifting facilities required.
4. Limitations on building design imposed by shape and size of panels.
5. Difficulties in making connections between panels.

The disadvantages result in high costs and prefabrication has not become common although it has not entirely died out. It is likely that it will find a niche in future for the production of elements of limited size. An example of off-site, prefabrication of brick-work piers is shown in Fig. 6.12 illustrating the practical possibilities of this form of construction.

6.3.2 Alternative units

The continued use and popularity of brickwork would seem to depend mainly on appearance but supported by many favourable characteristics including durability, resistance to rain penetration, fire protection and sound insulation. It is also flexible in application and competitive in cost as compared with possible alternatives. It is undoubtedly because of these advantages that brickwork has remained the predominant masonry material for external walling notwithstanding the disadvantages of requiring a high input of skilled labour and a rather slow rate of construction. However, when used for cladding a steel or concrete building, where the masonry can be built after completion of the main structure, this limitation is less significant. Cavity walls are now almost always built with a blockwork inner leaf, seeking to benefit from the higher rate of building possible with larger units together with improved thermal properties. Blockwork is also widely used for external walls for which high-quality facing units are available and also for internal walls with plaster or dry finishes.

Over the years, attempts have been made to introduce new types of clay units which would result in higher productivity on site whilst retaining the advantages of the brick.

Figure 6.12 Construction of prefabricated brickwork piers. (Courtesy of The Brick Development Association and photographer Graham Gaunt.)

One such unit was the V-brick (Fig. 6.13(a)), developed at the Building Research Station in the late 1950s and later marketed by Redland. The idea of the V-brick was to permit the construction of a cavity wall in a single operation. To this end, it consisted essentially of a pair of perforated clay bricks interconnected by thin webs. A 178 mm thick solid brick, known as the Calculon (Fig. 6.13(b)), was also produced as a substitute for the conventional 225 mm 'one brick' wall and although both were successfully used in multi-storey buildings, they were not widely adopted and are no longer produced. The principle of these units is sound and the lack of success of the V-brick probably resulted from increasing requirements for thermal insulation. The Calculon was primarily aimed at high-rise construction, which more or less came to a stop in the UK shortly after this unit was offered.

Also in the 1960s, the British Ceramic Research Association developed large hollow block units known as MG blocks (Fig. 6.13(c)) intended for use in low rise buildings. The blocks were 300 or 400 mm in height and width and 100 mm thick and of quite robust section with 50% perforations to reduce weight. Various jointing systems were envisaged and it was possible to assemble the blocks to form storey height planks. Storey height units were also produced which could be joined to give panels. The feasibility of the concept was demonstrated but was not carried into commercial production. A storey height unit was also developed in France some years later. The cross-section was cellular and was produced in thicknesses of 300 and 200 mm

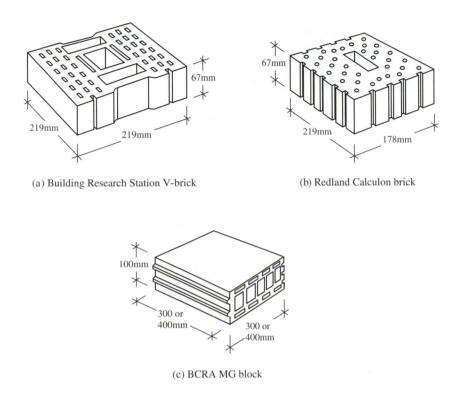

(a) Building Research Station V-brick   (b) Redland Calculon brick

(c) BCRA MG block

Figure 6.13 Non-conventional clay block units tried in the past.

and in widths of 600 and 300 mm. The makers claimed that construction using these units had advantages in cost, construction time, dimensional stability, thermal characteristics and flexibility in design over conventional brickwork. These units would have been technically difficult to produce and therefore expensive which may account for their not having found a place on the market. Nevertheless, in exploring the possibilities of introducing new and improved units it is important to examine earlier attempts and to discover why they did not succeed.

In recent years considerable attention has been devoted to the development of large blocks in various materials. Very large, solid blocks of lightweight material including, aerated autoclaved concrete (AAC or Aircrete) have been produced with a view to obtaining improved thermal properties and increased productivity on site. In addition to large solid blocks, a variety of special units have been produced in Europe, including JUWÖ Poroton and KLB blocks, both manufactured in Germany. Poroton units, illustrated in Fig. 6.14, are cellular, available in sizes up to $490 \times 300 \times 238$ mm. They are made of lightweight material produced by mixing small polystyrene beads with the clay before extrusion and firing to give enhanced thermal properties. One type of block by this manufacturer, designated the 'P-brick', is ground on the bed and top surfaces which is claimed to reduce construction time by 35% along with other advantages which result from the use of a thin mortar joint. The method of laying large Poroton blocks is illustrated in Fig. 6.15.

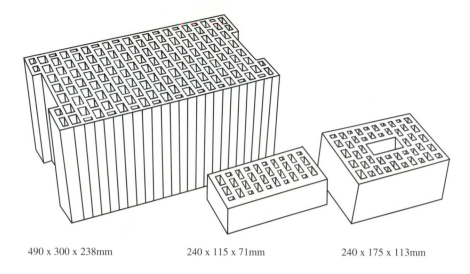

490 x 300 x 238mm   240 x 115 x 71mm   240 x 175 x 113mm

Figure 6.14 JUWÖ Poroton clay blocks.

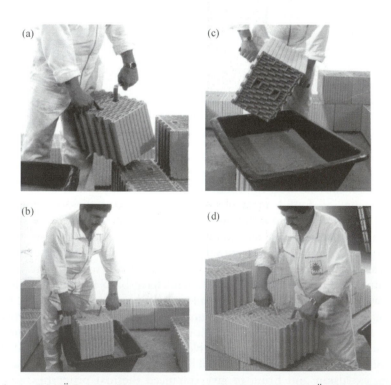

Figure 6.15 Laying JUWÖ Poroton Planziegel system blocks. (Courtesy JUWÖ Poroton-Werke (http://
www.juwoe.de).) (a) Block lifted from stack with special gripping tool. Note the keys and
grooves on the ends of the blocks. (b) Block dipped in mortar container. A thin bed mortar
is used with the accurately sized blocks. (c) The mortar adheres to the clay. (d) The block is
placed on the wall.

Figure 6.16 A light crane on a small building site. Inset shows the safe load/radius relationship for this crane.

A range of KLB blocks is produced for different functions, including special units for basement construction and others, with enhanced thermal properties, for outer walls. Also available are very large blocks, a type with ground surfaces and a system designed to be laid without mortar. All are made in a range of sizes.

The increased use of large blocks has implications for site practice having regard to weight limits on handling them. Thus for one-handed laying the weight must not exceed 3.6 kg and units of about 8 kg require to have grip holes. Beyond this, units have to be laid with two hands, and very large blocks, exceeding about 20 kg would require mechanical handling equipment on site in the form of a light crane such as that shown in Fig. 6.16. Figure 6.17 illustrates the laying of KLB blocks in this way, using thin bed mortar with a mortar sledge. Related to the size of block is the amount of mortar used in a given area of masonry: clearly, this decreases with increase in size of unit. Furthermore, very accurately manufactured units make it possible to have joints between 1 and 3 mm in thickness, provided that suitable thin bed mortars are used. These have a maximum particle size of 1 mm and are laid with a special trowel or mortar sledge (Fig. 6.17). Thin bed mortars have the advantage of setting quickly so that the masonry can be taken to a greater height in a shorter time after laying as compared with conventional practice. A site trial comparing productivity as between

Figure 6.17 Laying KLB blocks using a crane. Note thin layer mortar bed laid by mortar sledge. (Courtesy KLB Klimatleichtblock GmBH (http://www.klb.de).)

conventional bricklaying and large size blockwork laid in thin bed mortar has suggested that using the latter technique about double the square meterage per man-day could be achieved.

The logical end point in thin bed mortar construction might be to have no mortar between the units and, as mentioned above, such systems are available. Omission of the mortaring operation must obviously lead to great saving of construction time but the success of such a method of building must depend on preventing rain penetration through the wall. This could be achieved by external plastering or where mortarless masonry is used in cavity wall construction or as cladding to a concrete wall.

6.3.3 Site practice

Changes in site procedures will of course follow the introduction of prefabrication or the adoption of new types of masonry unit both of which imply the development of skills different from those required for the construction of traditional style masonry. A demand will, however, remain for highly skilled masons where high quality facing brickwork is required, where there is intricate detailing and for the repair and restoration of old buildings. It is also to be expected that with the greater accuracy in assembly, which these new methods imply, a greater degree of supervision will be necessary and therefore a need for more highly skilled supervisory staff.

More rapid methods of masonry construction will result in reduction of time lost on site due to adverse weather and in generally improved conditions for operatives. Further improvement in site conditions may be achieved by the use of protective shelters over the building under construction to create at least an approach to a factory environment.

6.3.4 Masonry building systems

The use of conventional masonry units as components of a system for constructing the shell of a building has been explored in the past in applying prefabricated construction. In a different way, a system was developed by the British Ceramic Research Association in the 1980s using standard sized bricks and taken as far as the construction of prototype houses. The system, known as SLIM (Single Leaf Insulated Masonry), made use of 102.5 mm clay brickwork walls with 50 mm 'Styrofoam' internal insulation and 12.7 mm plasterboard, giving a calculated U value of $0.5 \, \text{W/m}^2\text{K}$. A typical section of the construction is shown in Fig. 6.18. Although thoroughly tested in relation to structure, rain penetration and detailed design, the system was not adopted by the building industry, perhaps because in certain respects it is rather close to the already well established timber frame construction method although making structural use of the brick cladding which is in any case present. It is, however, an excellent example of the systematic development of a masonry building system.

As already described, a number of block manufacturers in Europe offer their product in a range of formats from which various types of walling can be constructed. Examples include Poroton blocks that can be supplied in a comprehensive range of sizes and types along with subsidiary materials from which solid and cavity walls can be assembled. KLB blocks are similarly available in many types for different functions, along with lintels, floor elements and other components. In both cases, a complete system for the shell of the building is offered greatly simplifying logistical problems.

In the Netherlands, a system of building in calcium silicate blocks is in use whereby the set of blocks required to build each wall of a house is first worked out on computer and delivered as a set to site, together with the necessary thin bed mortar materials. Large blocks are used and laying is carried out by a team of two masons using a small crane. This system has proved very successful in practice and is widely used in that country.

6.3.5 Concluding comment

It is clear that masonry construction has undergone considerable change through the 20th century with the virtual disappearance of stone masonry for the construction of important buildings, apart from the use of thin slabs as cladding. Similarly, in the UK, clay brickwork has been largely replaced for the inner leaf of cavity walls by blockwork in different materials, which has also found much wider general application.

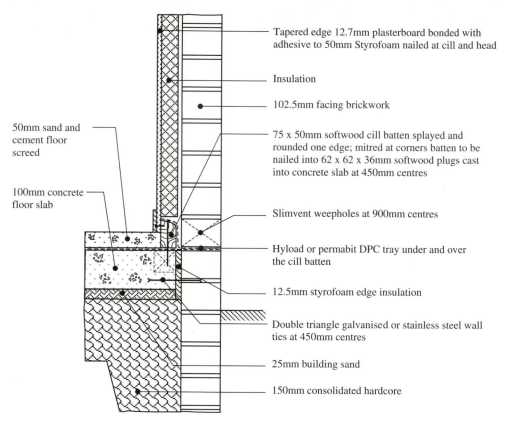

Tapered edge 12.7mm plasterboard bonded with
adhesive to 50mm Styrofoam nailed at cill and head

Insulation

102.5mm facing brickwork

75 x 50mm softwood cill batten splayed and
rounded one edge; mitred at corners batten to be
nailed into 62 x 62 x 36mm softwood plugs cast
into concrete slab at 450mm centres

50mm sand and
cement floor
screed

100mm concrete
floor slab

Slimvent weepholes at 900mm centres

Hyload or permabit DPC tray under and over
the cill batten

12.5mm styrofoam edge insulation

Double triangle galvanised or stainless steel wall
ties at 450mm centres

25mm building sand

150mm consolidated hardcore

Figure 6.18 Cross section of Single Leaf Insulated Masonry (SLIM) wall developed by British Ceramic
Research Association. (Courtesy CERAM Building Technology (http://www.ceram.co.uk).)

Since the late 1960s thermal insulation has become an increasingly important factor in
the design of external walls and this has led to the adoption of blocks having good
insulating properties and to the incorporation of separate insulating materials. Larger
sized blocks have been introduced together with thin bed mortars with a view to
increased productivity and some attention is being given to improving site conditions
for building operatives.

   Prefabrication of brickwork wall panels has been attempted in several countries but
has not so far been widely adopted. It would, however, appear promising if used for
elements of limited size. Various systems have been developed in recent years based on
the use of large blocks and thin bed mortars or in some cases mortarless construction.
Adoption of these systems implies considerable changes in logistical arrangements
whereby manufacturers or suppliers will have to become more closely involved in
ensuring that the appropriate materials arrive on site as and when they are to be
used. Site procedures will also change with the use of more mechanical lifting appli-
ances and improved protection against adverse weather conditions.

Further reading

Brickwork: good site practice, Knight, T., Brick Development Association, BDA Publ. BN 1, 1991.

Workmanship on building sites, Code of Practice for Masonry, British Standards Institution, BS 8000: Part 3: 1989.

Chapter 7

Defects in masonry walls

7.1 General

Although masonry walls are durable and normally capable of fulfilling their function for many years with very little maintenance, defects are sometimes encountered, usually becoming apparent as cracks or by rain penetration. These can arise from deficiencies in design or construction or from service conditions or a combination of these. Some defects may show up soon after completion of a building whilst others may develop after many years. In some circumstances no symptoms are apparent to the user but an inherent defect may have the effect of reducing the strength of the masonry possibly with eventually serious consequences.

A system of classification of visible damage to walls has been suggested by the Building Research Establishment, (BRE Digest 251), initially under the headings 'aesthetic', 'serviceability' and 'stability'. These are almost self explanatory, referring first to appearance, secondly to functional defects such as weathertightness and finally to structural safety. These descriptions are categorised on a scale from 1 to 5 and related to crack widths from 0.1 mm to 25 mm. Damage resulting from moisture and thermal movements is stated rarely to exceed the third category on this scale and in low rise buildings, to which it applies, only minor repairs would be required. On the other hand, such damage in a high rise building could give rise to very expensive repair work, for example in replacing an entire leaf of masonry. It is important, however, not to judge the situation entirely on crack width since serious defects, such as the corrosion or absence of wall ties, can be revealed initially by the appearance of quite narrow cracks.

Avoidance of defects clearly begins at the design stage and one of the first and most important decisions having implications for satisfactory service behaviour is the selection of materials. From the foregoing chapters it will be understood that this depends on structural, thermal and acoustic requirements as well as on durability, having regard to environmental conditions. Having decided on the layout of the building and the masonry materials to be used there follows the equally important stage of detailed design. This will include consideration of damp-proof courses, cavity trays, fire stops, the location of movement joints, services and other essential matters.

The designer's requirements and intentions must be conveyed to the builder in drawings and specifications, referring where necessary to standards and codes of practice. Details should be kept as simple as possible, attention being given to the avoidance of ambiguity or simply errors in the information supplied to site.

Alterations are frequently made in the course of the design of a building and it is important to see that consequent amendments are made to all drawings and documents affected. There is sometimes a temptation for a builder to circumvent errors of this kind in a makeshift manner which results in trouble later on.

7.2 Cracking in masonry buildings

Apart from overstressing, the appearance of cracks in masonry is the result of movement beyond that which can be accommodated by the material. The causes of movement have been discussed in Chapter 5 with indications as to how allowance should be made to accommodate those resulting from temperature and moisture change. It was also noted that foundations must be designed to avoid excessive movement and that selection of materials must be such as to minimise the risk of unwanted chemical reactions taking place.

If cracking has taken place, identification of the cause is an essential preliminary to any decision on remedial work. The causes of cracking are often complex and their diagnosis requires considerable experience as well as an understanding of the likely physical processes involved.

In inspecting cracks the following features should be observed:

1. Direction, extent and pattern.
2. Width, noting tapering and relation to masonry courses.
3. Depth.
4. Whether edges are sharp (indicative of tensile failure).
5. Indications of age (e.g. whether clean, painted, etc.).

This information should be plotted on sketches of the walls along with details of construction and carefully studied to ascertain the direction of movement and thus possible causes. If there are indications that movement is still taking place, it may be necessary to monitor the behaviour of the building for some time.

Figures 7.1–7.4 show some illustrative examples of crack formations associated with commonly encountered causes. Figure 7.1(a) relates to thermal expansion of the masonry in a parapet and 'oversailing' at damp-proof course level. The other examples in Fig. 7.1 illustrate typical crack patterns resulting from moisture expansion in clay brickwork.

Calcium silicate and concrete masonry is subject to shrinkage after construction and may display cracking as in Fig. 7.2. It will be noted that the units in these forms of masonry are normally set in relatively weak mortars (1:1:6 cement:lime:sand or lower designations) so that cracks will tend to follow the joints rather than pass through the bricks or blocks which, however, will happen if a strong mortar has been used. Shrinkage cracks usually begin at an opening or change of section and tend to be vertical with about the same width from top to bottom.

Typical cracking resulting from soil movement is shown in Fig. 7.3. A crack tapering from top to bottom or vice versa, depending on whether the soil movement is heave or settlement can be produced in a long wall (Fig. 7.3(a)). If there are openings in the wall cracks will originate at corners, their direction depending on whether they result from settlement or heave (Fig. 7.3(b)). Settlement cracks are often found at the

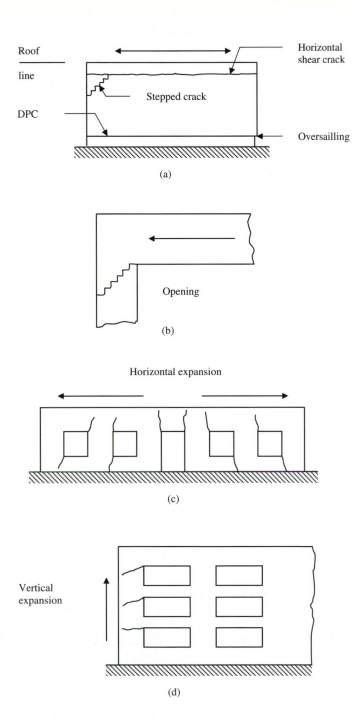

Figure 7.1 Cracks developed as a result of masonry expansion.

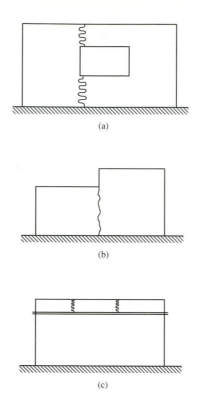

Figure 7.2 Shrinkage effects in calcium silicate brickwork or concrete blockwork.

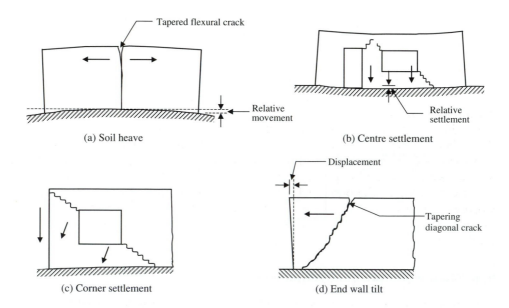

Figure 7.3 Evidence of soil movement.

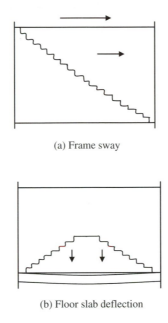

(a) Frame sway

(b) Floor slab deflection

Figure 7.4 Cracking due to deflection of structural frame.

end of a building resulting from corner settlement as in Fig. 7.3(c) or tilting of an end wall as in Fig. 7.3(d). Soil movement or foundation settlement may result in large cracks requiring expensive remedial work although the load bearing capacity of a wall may not be seriously reduced provided that the wall affected is not leaning or bulging.

Cracking of walls can also result from the deformation of a surrounding structural frame, examples being shown in Fig. 7.4. Brickwork cladding is sometimes supported between the floor slabs of a concrete building (Fig. 7.5) shrinkage of which may cause buckling of the wall and detachment of the slips covering the edge of the slab. Although this detail is avoided in current practice, it is still known to cause trouble in older buildings constituting a potential danger to people below and requiring very expensive repair work.

In addition to the causes described above, cracks can result from overstressing in the vicinity of beam or lintel bearings or from wall tie deficiencies in cavity walls. The latter may be the omission of the requisite number of ties, which may show up as horizontal cracking or in more extreme cases by bulging of one of the leaves. Complete failure of the outer leaf may occur when subject to strong wind suction. Corrosion of galvanised wall ties is not uncommon, especially in marine environments, and may become apparent through local horizontal cracks and rust stains on the outside of the wall.

7.3   Rain and damp penetration

Resistance of masonry to rain penetration depends on a number of factors including the characteristics of the units and mortar, surface treatment and very importantly on

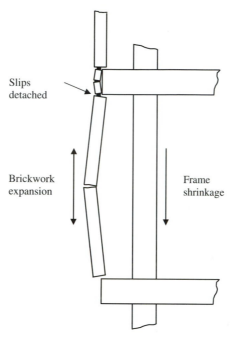

Figure 7.5 Infill masonry panel damaged by differential movement between wall and frame.

the skill of the mason. Traditionally built walls of considerable thickness using absorptive materials are effective in keeping rain out of a building, even in severe conditions of exposure, as water will be held in the wall and evaporated off in periods of dry weather. Modern construction, however, being very much thinner, is more critically dependent on good workmanship as badly filled joints provide a ready path for rain to pass through the wall. Perpend joints are particularly vulnerable as are joints between relatively impermeable units through which water can be drawn by capillary action.

The outer leaf of a cavity wall, or thin masonry cladding, has mainly the function of a rain screen and, particularly in areas subject to driving rain, are likely to be penetrated by rain and to remain wet for prolonged periods. It follows that failure to install damp-proof courses and cavity trays properly will result in water finding its way into the building. A very common defect arises when cavities are bridged by mortar or debris which has been allowed to fall into them in course of construction. Also, if the drains at the base of the cavity are omitted, placed at the wrong level or blocked, water is likely to accumulate above the cavity tray and eventually find its way into the building.

The appearance of moisture, possibly in considerable quantity, on the inside of a wall is sometimes mistaken for rain penetration whereas the cause may be condensation. Rain penetration tends to occur in patches with well-defined edges and to fade in periods of dry weather. Condensation on the other hand appears on cold walls and in ill ventilated spaces, especially apparent on impervious surfaces. Insulation of the wall may be inadequate or unsuitable but condensation depends very much on the use of the building, particularly on standards of heating and ventilation, and is not specific to masonry construction.

Moisture penetration may occur at ground level as the result of a faulty or omitted damp-proof course or, in the case of a solid wall, by bridging of the damp-proof course by mortar or by earth or paving lying against the outside.

Omission or faulty installation of a sub-floor damp-proof membrane where it abuts a wall is also a possible cause of internal dampness. At roof level, water may penetrate a wall if the damp-proof course under a parapet or chimney-stack is faulty. Leakage may also be experienced between flat roof coverings and walls or at flashings where a pitched roof joins a wall.

Poor detailing at window and door openings can result in the surrounding areas of wall becoming damp, particularly in exposed sites. It should be noted that wind pressure can result in a static head of possibly 150 mm, forcing rain water through leaky joints and overcoming an upstand in a cavity tray of less than this height.

7.4   Other visible defects

A number of chemical and physical effects, which give rise to visible defects, have been mentioned in Chapter 5 including frost action, salt crystallisation and sulphate attack. Frost action typically causes expansion of joints and delamination and is most likely to be found where the material is persistently wet and subject to repeated freeze/thaw cycles. Thus exposed locations such as parapets, garden walls and the outer leaves of heavily insulated cavity walls are particularly vulnerable.

Surface crystallisation (efflorescence) is common in newly built masonry as the result of the evaporation of water containing dissolved sodium or potassium salts. This gives rise to the deposition of white crystals, which may be general or localised in the vicinity of a detail, and this encourages prolonged retention of water in the masonry. Such efflorescence is unsightly but in most cases is subsequently washed off by rainwater. In this case it is not a permanent problem although it can recur over a considerable period and may have to be removed by wire brushing. More persistent staining can be caused by leaching of lime or gypsum from cement or concrete as in the example shown in Fig. 7.6. Insoluble salt stains constitute a more serious problem and may have to be removed by chemical treatment.

As explained in Chapter 5, sub-surface crystallisation can produce delamination or spalling, similar in appearance to frost damage. This is uncommon in brickwork but in stone masonry can be extensive giving rise to the term 'stone decay'. Serious damage to stone masonry can also be caused by the use of unsuitably strong cement mortar for pointing, as shown in Fig. 7.7.

Sulphate attack of masonry is indicated by the appearance of cracks following the bed joints resulting from expansion of the mortar by the products of the chemical reaction. The mortar becomes lighter in colour and eventually disintegrates leading to the complete disruption of the masonry. In its earlier stages sulphate attack may be mistaken for frost damage from which it can be distinguished by chemical analysis of the mortar. Cracking along the bed joints can also result from corrosion of wall ties in cavity walls but in this case the pattern of cracks may indicate the presence of the ties and rust stains may be visible. If a wall is rendered, sulphate attack of the masonry

Figure 7.6 Lime staining on a concrete blockwork wall.

Figure 7.7 Stone masonry wall severely damaged by the use of unsuitably strong mortar for pointing.

may lead to the appearance of numerous horizontal cracks caused by the expansion of the mortar.

7.5   Workmanship factors affecting strength

It is a commonly held view that masonry is a rather weak, variable material compared to steel or concrete. This has arisen from the fact that in the past it has been thought of as a 'building' rather than an 'engineering' material, meaning that construction has been left in the hands of workmen of uncertain skill and supervision has tended to be minimal. This need not be the case and a vital step in producing structural masonry of consistent strength and quality is for masons and supervisors to be aware of the workmanship factors which are important in this respect. It will be understood that we are not here concerned with gross errors such as the use of the wrong bricks or mortar or with the variability of materials as such but with the identification of factors in site work which can affect the strength of masonry. These include the following:

1.   Incorrect proportioning and mixing of mortar.
2.   Incorrect adjustment of suction rate of masonry units.
3.   Incorrect jointing procedures.
4.   Disturbance of units after laying.
5.   Failure to build walls 'plumb and true to line and level'.
6.   Failure to protect new work from the weather.

In discussing the strength of masonry in Chapter 4, it was shown that the strength of mortar, as defined by the cube crushing strength, is not a very critical factor—for example, with bricks of crushing strength $35 \, \text{N/mm}^2$ a halving of mortar cube strength from $14 \, \text{N/mm}^2$ to $7 \, \text{N/mm}^2$ may be expected to reduce the compressive strength of the brickwork from about $16 \, \text{N/mm}^2$ to $14 \, \text{N/mm}^2$. This corresponds roughly to a change in mortar mix from $1:3$ cement : sand to $1:4\frac{1}{2}$ or say 30% too little cement in the mix. A similar reduction in mortar strength could be brought about by an excess of water—moving from a water/cement ratio of about 0.6 to 0.8 in a typical case.

Variation in mortar strength of the order mentioned above has a greater influence on the flexural strength of clay brickwork, a change from $1:3$ to $1:4\frac{1}{2}$ cement : sand reducing the strength by some 25% according to BS 5628: Part 1. The effect, however, is considerably less in the case of calcium silicate and concrete bricks and blocks.

In adjusting the water content of mortar to achieve workability, it is necessary to have regard to the amount of water that will be abstracted from the mortar when the unit is laid. This is defined by the suction rate or initial rate of absorption (measured in $\text{kg/m}^2/\text{min}$). Excessive removal of water may leave cavities in the mortar which fill with air and result in a weakened material on setting. On the other hand, brickwork built with saturated bricks develops poor adhesion between brick and mortar and is liable to frost damage and other troubles. Some specifications recommend a limiting suction rate or alternatively the use of a high retentivity, lime mortar to control the extraction of water. In some cases it is desirable to dip (but never soak) the brick in water before laying.

The proper filling of bed joints is critical in relation to the compressive strength of masonry and may result simply from failure to spread the mortar across the full width of the unit or from the practice of 'furrowing' whereby a gap is deliberately created in the centre of the mortar bed. This eases the mason's task by reducing the tendency for mortar to be extruded when the next unit is placed on it. Experiments have shown that incomplete filling of the mortar bed may reduce the compressive strength of the masonry by as much as 25% to 30%.

Failure to fill the vertical joints between units does not have much effect on the compressive strength but is likely to reduce the flexural and shear strength and to adversely affect sound insulation and resistance to rain penetration.

The thickness of the bed joints in masonry relative to the height of the unit is of particular significance in brickwork but less in blockwork. This is because compressive failure tends to result from lateral expansion of the mortar being restrained by the unit, inducing tensile stress in the latter. The thicker the mortar joints in relation to the height of the unit, the higher will be these tensile stresses. Thus an increase in the thickness of the bed joint from the normal 10 mm to 16 mm can result in a 25% reduction in strength.

Another laying defect arises from the practice of spreading too long a bed of mortar—only sufficient should be spread as will permit the units to be set in plastic mortar. Somewhat related to this is the breaking of the bond between unit and mortar by disturbance after laying with possible reduction of strength and resistance to rain penetration.

An essential requirement in masonry wall construction is to ensure that walls are built 'plumb and true to line and level'. It is clear that departures from this requirement are likely to reduce the compressive strength of a wall by increasing the eccentricity of loading. Three possible faults can be distinguished; off plumb, bowed and out of alignment with walls above and below. Inaccuracies of 15 mm to 20 mm in a storey height have been found in practical construction and this could result in strength reductions of about 15%.

The ambient temperature and humidity in which a wall is located after construction may affect its final strength. As noted in Chapter 6 freezing of the mortar before it hardens will lead to reduced strength although experiments in Finland showed little or no reduction in strength of brickwork built in temperatures as low as $-15°C$. It is well known, however, that walls built under freezing conditions are liable to develop undesirable deformations which could give rise to reduced compressive strength. Masonry built in hot, dry conditions are likely to lose water from the mortar, which could prevent complete hydration of the cement and thus reduced strength and adhesion. Heavy rain could also have an adverse effect. It is obviously necessary to protect newly constructed masonry against extreme weather conditions, whether cold, hot or wet.

The separate effect of a number of workmanship factors has been discussed. In any particular case, these defects will be present in varying degrees and the overall strength of a wall will reflect their combined effect. An Australian report makes the following assessment of the various defects in terms of reduction in strength of a wall built under laboratory conditions:

1. Furrowed bed = 10%
2. 16 mm bed joints = 25%
3. Perpend joints unfilled = Nil

4.   12 mm bow = 15%
5.   Outside cure (warm conditions) = 10%

It was considered that these effects were non-interactive and additive so that in an extreme case the reduction in strength could be as much as 60%. This is consistent with the similar tests conducted in the USA. It has to be said that a wall built as badly as the worst in these tests would (or should) be regarded as unacceptable in a practical situation.

These workmanship factors may therefore considerably reduce the site strength of masonry but in most cases defects will not be apparent without detailed investigation. It was explained in Chapter 4 that the difference in strength of masonry between that assumed in design and that attained on site is allowed for by dividing the characteristic strength by a partial safety factor. Thus in Eurocode 6, the partial safety factors for the lowest defined standard of workmanship are increased by 40 to 50% as compared with the highest. This is a reasonable reflection of the possible strength reduction resulting from the workmanship defects referred to above.

Further reading

Hinks, J. and Cook, G. (1997) *The Technology of Building Defects*, E. & F. N. Spon, London.
Richardson, B. A. (1991) *Defects and Deterioration in Buildings*, E. & F. N. Spon, London.
Further observations on the design of brickwork cladding to multi-story RC frame structures, Foster, D., Brick Development Association, BDA Technical Note 9, 1975.
Low rise buildings on shrinkable clay, Building Research Establishment, BRE Digests 240–242, 1980.
Assessment of damage in low-rise buildings, Building Research Establishment, BRE Digest 251, 1990.
Common defects in low-rise traditional housing, Building Research Establishment, BRE Digest 268.
Why do buildings crack? BRE Digest 361, 1991.
Rising damp in walls: diagnosis and treatment, Building Research Establishment, BRE Digest 245, 1981.

Chapter 8

Repairs and improvements to masonry walls

8.1 Maintenance and repair

8.1.1 Pointing

Normal maintenance of masonry walls, assuming that there are no defects such as bulging or cracking, consists of pointing the joints—at intervals of decades rather than years. The wall must first be carefully prepared by removing loose mortar, raking out the joints to a depth of 12 to 15 mm and removing loose fragments and dust. If there is loose mortar to a greater depth caution should be exercised over the area to be raked—perhaps to about three courses of brickwork over a length of three stretchers. Before placing the mortar the joints should be wetted to prevent undue abstraction of water.

The mortar used for pointing must be chosen to be compatible with the original masonry in terms of strength and colour. On no account must a strong cement mortar be used with low to medium strength bricks or softer natural stones as this may lead to spalling as shown in Fig. 7.7. A 1 : 1 : 6 cement : lime : sand mix is generally suitable for modern brickwork and 1 : 2 : 9 for softer brickwork and natural stone. In general, the mortar should contain only enough cement to ensure durability and well graded, washed sand without fines used so as to avoid shrinkage. A high water content should be avoided, if necessary by using a limited amount of air-entraining liquid. Colouring agents are available and should be used in accordance with manufacturers instructions. Some experimentation may be necessary to achieve the required colour but not more than 10% of such pigment or 3% of carbon black, should be added.

Buildings over 100 years old will usually have been built using lime mortar which should also be used for pointing. Specialist advice should be sought in selecting a mortar mix for old buildings as serious damage can result from the use of cement.

The finish of the pointing will be decided according to the desired appearance but for good weather resistance the profile should be as in Fig. 8.1(a), (b) or (c). Buttering over the joint, which is often done on rubble stonework, carries the danger of water being conducted into the masonry if the mortar shrinks and is therefore inadvisable. Recessed joints as in Fig. 8.1(d) should be restricted to sheltered locations.

8.1.2 Non-structural repairs

Repairs beyond routine maintenance by pointing may be required to rectify defects of the kind described in Chapter 7, but which are not of structural significance.

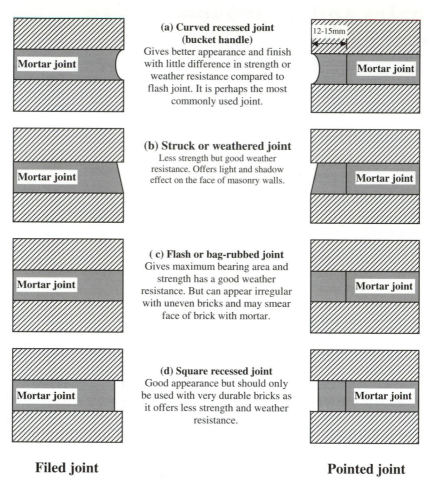

Filed joint   **Pointed joint**

Figure 8.1 Types of mortar joint finish used for masonry construction.

If cracks resulting from thermal or moisture movement have been identified, before carrying out repairs it may be advisable to monitor the situation for a time to ascertain whether the movement has stopped. If it has not, the repair should be regarded as temporary.

Cracks less than about 1.5 mm generally will not require to be filled. Slightly wider cracks, especially in exposed walls built in non-absorbent materials, may be repaired with special cementitious or resin compounds. Wide cracks following bed and perpend joints can be raked out, filled and pointed with normal mortar to match the undamaged masonry. If cracks pass through units these may require to be cut out and replaced.

Frost damaged bricks will also require to be replaced or the faces replaced by brick slips. If in the outer leaf of a cavity wall such damage is extensive, the question will arise as to whether the units used are so unsuitable that the wall has to be rebuilt in materials appropriate to the exposure.

Appearance of repaired work is frequently a problem owing to the difficulty of obtaining supplies of the same materials some years after construction. Re-pointing or even rebuilding of a section of a wall greater than the damaged area may be required in order to ensure an acceptable result. Alternatively, rendering or cladding of a frost damaged wall may be a solution in certain circumstances.

Surface damage in stone masonry may be treated by redressing the stone and re-pointing but if stones are damaged to a considerable depth, perhaps over 20 mm, it may be necessary to build up the affected area with mortar or, more satisfactorily, replace it with natural or artificial stone. In either case it will usually be difficult to achieve a close colour match; some conservationists hold the view that it is acceptable to show where such restoration has been carried out.

Parapet walls and freestanding walls are exposed to driving rain and frost on both sides and thus likely to be more frequently in need of repair than the exterior walls of occupied buildings. Frost damage, resulting in spalling of bricks, erosion of mortar and cracking or bowing are common defects. Assuming that the damage has not made the wall unstable, repair may range from re-pointing to reconstruction in more resistant materials in the case of frost damage or the provision of movement joints. Defective parapet walls are sometimes responsible for giving rise to water penetration into a building. In such a case, improved copings, damp-proof courses and flashings may have to be provided (Fig. 8.2).

Chimneystacks and walls containing flues are also prone to damage from exposure to driving rain and frost. The presence of sulphates in flue gases can have a destructive effect on mortar in the presence of water derived either from rain penetration or condensation. Brickwork chimneystacks often develop a bow or lean away from the direction of the prevailing wind as a result of sulphate attack which is greater on this side. Disused flues and chimneystacks are also vulnerable and should be capped and ventilated to prevent deterioration. Again repairs range from re-pointing to rebuilding using appropriate materials.

As described in Chapter 7, the appearance of damp on the inside of a masonry wall can result from the faulty installation or omission of a damp-proof course or cavity tray or from mortar droppings in a cavity. Investigation of such a defect is carried out with the use of a 'borescope'—a kind of miniature periscope which can be inserted through a small hole in the masonry. This will reveal the presence of mortar droppings on wall ties and at the bottom of the cavity (and also the absence or faulty installation of wall ties). It will probably be necessary to remove some units to reveal the full extent of the problem, especially where damp-proofing material has been incorrectly placed or omitted. If the extent of damp penetration is limited, repair can be effected by removal of a few units but if extensive, for example by the omission of a cavity tray, it may be necessary to remove several bricks or blocks at a time along the length of a wall. In extreme cases, it may be necessary to demolish and rebuild the outer leaf of a cavity wall.

In old buildings it may be found that there is no damp-proof course and to eliminate rising damp it will be necessary to install one. In brickwork walls this can be done by cutting a slot through the thickness of the wall, in 1-m lengths, and inserting a bituminous DPC. In stone masonry rising damp can be controlled by the injection of silicone or aluminium stearate through holes drilled in the wall from the outside.

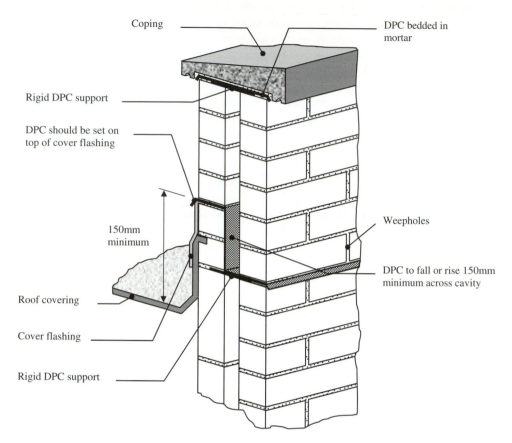

Figure 8.2 Typical detail of flashing and DPC in a parapet wall. (Based on BRE Digest 380.)

8.1.3   Structural repairs

Examination of a building may reveal defects which affect its structural integrity. These may be cracks of various kinds, bulging or leaning of walls, as described in Chapter 7.

Most cracks are not serious from this point of view but walls out of plumb by more than 25 mm or bulging by more than 10 mm would call for structural repairs as would reduced bearing area or cracking in the vicinity of lintel or beam ends. Defects in thin leaf cavity walls are more critical than those in thick walls. Thus a minor bulge or bed joint cracking in an outer leaf may be indicative of missing or corroded wall ties. Should this prove to be the case, it will be necessary to install special ties of which several types are available (Fig. 8.3) or even to rebuild the wall.

Leaning or bulging walls in old buildings may also have to be rebuilt or in some cases may be stabilised by the insertion of tie rods through the building (Fig. 8.4). Buttresses are occasionally used for this purpose but are only likely to be effective if they exert a force against the wall requiring support and are not subject to settlement, which could simply add to the problem. Even where walls are not significantly leaning

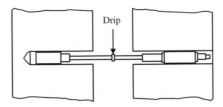

Wall tie with plastic expanders, stainless steel tie rod, nuts, etc.

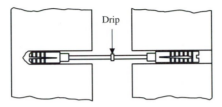

Wall tie with stainless steel expanders, tie rod, nuts, washers, etc.

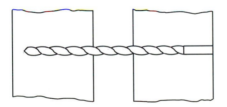

Self-tapping, hard copper alloy tie

Figure 8.3 Retrofitted wall ties of various types.

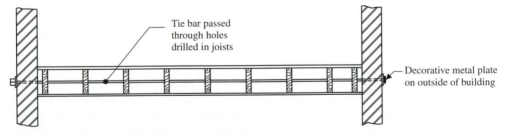

Figure 8.4 Stabilising walls by tie bars passed through width of building.

or bulging, it may be necessary to install ties between floors and walls to bring their stability into line with current standards. This can be achieved by the use of galvanised or stainless steel straps screwed to the flooring or by tie bars within the depth of the flooring. In the latter case the tie bars may be secured to at least three joists or may pass through the width of the building. Typical details are shown in Fig. 8.5.

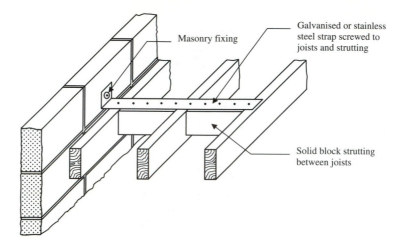

(a)   Steel strap to joists parallel to wall (screwed to joists if these are at right angles to wall)

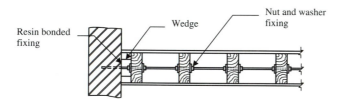

(b)   Steel bar clamped to joists and anchored to wall with polyester resin

Figure 8.5 Retrofitted lateral restraint to a masonry wall.

Stone masonry walls, likely to be of considerable age, require special attention as apparently sound walls may have internal voids or very weak core material, the original lime mortar having perished. Methods of repair include injection with cementitious or resin grout or stitching with stainless steel rods (Fig. 8.6). Such operations, however, are of a specialised nature and have to be undertaken with great care.

Cracking resulting from foundation movement may require underpinning and filling of cracks but it is beyond the scope of this book to discuss this operation. In certain cases it has been suggested that as an alternative to underpinning it is possible to control settlement cracking by the insertion of bed joint reinforcement. Where covering over the surface of the masonry is acceptable, it is possible to repair a wall by fixing a steel mesh to the face of the masonry and embedding it in concrete applied by a high pressure spray. Such treatment, however, has obvious disadvantages in terms of appearance.

A major repair to the brick cladding of multi-storey concrete structures is required where storey height panels have been damaged by shrinkage of the concrete frame or

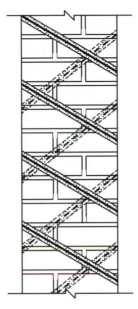

Figure 8.6 Strengthening of masonry wall by 'stitching' with stainless steel bars embedded in PFA/cement grout.

inner walls as in Fig. 7.5. In such cases it will usually be necessary to rebuild the brickwork with the insertion of soft joints at floor slab levels to permit differential movement to take place without causing further damage.

Another structural defect sometimes discovered in concrete or steel framed buildings is inadequate lateral strength of masonry cladding panels. This may arise from corroded or omitted ties to the columns, omission of support to the top edge of infill walls or the need for additional vertical support at mid-length of a panel. The detection of such defects is not easy in a completed building but where it is discovered remedial work is essential. Ties to columns similar to one of the types shown in Fig. 8.3 are readily retrofitted. If support to the top edge of an infill panel is required to develop sufficient resistance to wind forces, an appropriate repair would be the installation of a steel angle along the top of the wall, bolted to the soffit of the floor slab. This is generally a straightforward matter although it results in disruption of internal finishes. Retrofitting of vertical support, as indicated in Fig. 8.7, is possible in the form of additional concrete or hollow section steel columns but implies considerable disruption of interior finishes. Proprietary types of stainless steel windposts are available which can be installed by removing bricks from the outer leaf (Fig. 8.8).

8.2   Improvements and alterations

8.2.1   *Waterproofing of basements*

An improvement rather than a repair in older property is the installation of dampproofing in basements to permit their use as habitable accommodation. This can be

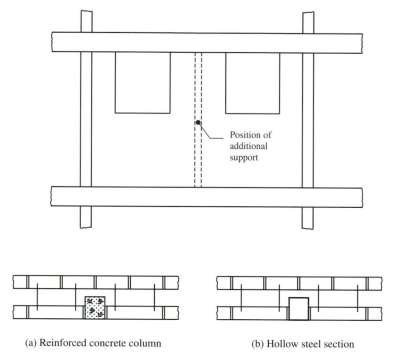

Position of
additional
support

(a) Reinforced concrete column   (b) Hollow steel section

Figure 8.7 Additional vertical support to laterally loaded panel.

carried out by several methods including the construction of an inner cavity wall, the application of asphaltic material to the wall or by dry lining over some form of membrane. Before proceeding with such work it is necessary to determine the cause of the dampness and the construction and condition of the walls to be treated. If a new inner cavity wall is constructed, the cavity has to be drained which may need the provision of a sump and automatically operated pump. There will be some reduction of the useable space with this solution and if this is unacceptable the alternative of dry lining may be preferred.

8.2.2 Formation of openings in walls

It is common practice in older, low rise properties to remove a loadbearing wall to form a larger room or to create a door opening. Needless to say, such an operation should only be attempted following advice from a structural engineer or other quali-fied person. Assuming that there are no signs of weakness or movement in the building, the procedure (Fig. 8.9) will be, first, to insert temporary support to the floor and wall above the opening being formed. This will be done by inserting a line of props and a spreader beam about a metre from the face of the wall to support the floor above. The wall above is supported by short beams or 'needles' inserted through holes in the masonry at the wall head, these needles being in turn carried by a second set of props. The opening in the masonry can then be formed and a beam or lintel inserted.

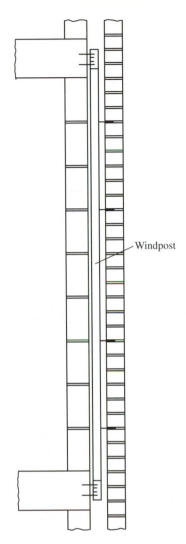

Figure 8.8 'Ancon' stainless steel windpost fitted in cavity.

The masonry is then built up between the needles and after allowing time for the mortar to harden, the needles are removed and the holes filled. Finally, the temporary supports for the floor are removed.

8.2.3 Improvement of thermal insulation

The thermal performance of masonry walls can be improved by the addition of insulation either internally, externally or in the case of cavity walls, by injecting insulating fill into the cavity. The latter procedure is straightforward and suitable materials are available as loose fills, which can be blown into the cavity, or as

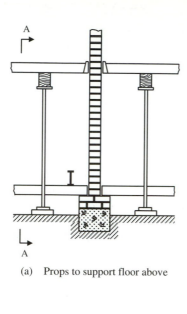

(a) Props to support floor above

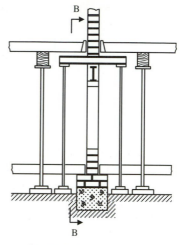

(b) Wall above supported by needles through wall

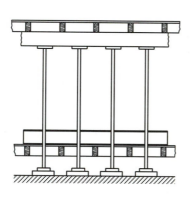

(c) Section AA showing floor supports about 1m from face of wall

(d) Section BB after formation of opening and placing of beam. Needles to be withdrawn and brickwork built up

Figure 8.9 Construction of opening in existing masonry wall.

foams which are injected. Loose fills include rock or glass fibre, polyurethane granules and expanded polystyrene. Polyurethane foam is injected as two liquids, which react to form a foam within the cavity. The technique is well established and is applicable to both new and existing construction and, if correctly installed has not been found to increase the risk of rain penetration. It is the most economical and least troublesome way of improving the insulation of existing cavity walls. The improvement in the U value achieved will of course depend on the characteristics of the fill and on the cavity width but in a small semi-detached or terrace house may be expected to reduce heating costs by some 25%.

If cavity insulation is not used, either by choice or because the building has solid walls, it is possible to improve insulation by applying it either to the inside or to the outside of the walls. Many considerations affect the choice between internal and external insulation, the most obvious including change of appearance of the building if external insulation is used and disruption of finishes, use of the building and loss of floor space if applied internally. If the building is intermittently heated (e.g. a hall or church) internal insulation has advantages in that less heat is absorbed by the wall so that the interior space will warm up more quickly. On the other hand, the wall will be colder than it was originally and therefore more susceptible to frost damage. Thought will also have to be given to the effect on any water pipes which may be outside the added insulation and to the possibility of condensation on the wall.

The application of internal insulation may be combined with the improvement of surface finishes either by using separate layers of plaster-board and insulating material nailed to battens on the wall or by using pre-bonded dry lining/insulating board fixed directly to the wall. Polystyrene and polyurethane insulation boards are available, the thickness depending on the required U value and the construction of the masonry wall but typically 50 mm.

Several methods are used for the installation of external insulation. If the building has a tile or weatherboard cladding, it may be possible to interpose a layer of rigid or flexible insulation without altering its appearance. Alternatively, a new outer cladding or rain-screen may be constructed in metal, glass or other material.

A ventilated and drained cavity should be provided between the external cladding and the insulation, as shown in Fig. 8.10(a), and if the insulation is combustible this cavity must be interrupted at intervals by fire resistant barrier.

For board insulation systems in which the insulation is protected by a rendered finish, materials such as polyurethane or mineral fibre, 20–100 mm thick may be used. These are fixed to the wall together with a sheet of expanded metal lath to provide a base for a 10–15 mm cement/sand render (Fig. 8.10(b)). Modifications will of course be necessary at openings to accommodate the thickness of the insulation. A somewhat similar method makes use of flexible insulation which can be taken round corners and into reveals without cutting or jointing.

8.2.4   Improvement of acoustic insulation

Improvement of the acoustic insulation of masonry walls is often carried out when an existing building is sub-divided to provide residential accommodation for different occupants and indeed will generally be necessary in these circumstances. As acoustic insulation is primarily dependent on mass, masonry walls are inherently good in this respect but, as explained in Chapter 6, their effectiveness may be reduced by flanking transmission. Thus, before attempting to carry out sound insulation improvements in old property, careful examination is essential to ensure that there are no concealed holes, for example at joist ends, between the occupancies, and that party walls extend to the full height of the roof space. Enhancement of the sound insulation properties of a wall can be effected by adding a leaf, preferably on both sides, with a 100 mm cavity between the new and existing walls. Additional leaves may be of brickwork or block-work, a 75 mm thick stud partition with glass fibre quilt between the studs or a proprietary form of partitioning. Such additional walls should be built from floor

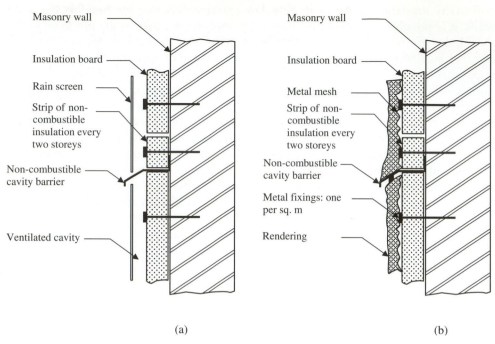

Figure 8.10 External board insulation on a masonry wall. (a) With rain screen and ventilated cavity. (b) With metal mesh reinforced rendering secured to the insulation.

to ceiling without attachment to the wall whose insulating value is to be augmented. Loss of floor space may be a limiting factor in the upgrading of acoustic insulation in this way and some less effective compromise may have to be accepted.

Further reading

Sowden, A. M. (1990) *The Maintenance of Brick and Stone Masonry Structures*, E. & F. N. Spon, London.

Repointing of brickwork, Harding and Bevis, Brick Development Association, BDA Publ. BN 7, 1976.

Repointing external brickwork walls, Building Research Establishment, BRE Good Repair Guide 24, 1999.

Repairing frost damage, Building Research Establishment, BRE Good Repair Guide 20, 1998.

Repairing chimneys and parapets, Building Research Establishment, BRE Good Repair Guide 15, 1998.

Repair and maintenance of brickwork, Brick Development Association, BDA Publ. TIP 1.

Installing wall ties in existing construction, Building Research Establishment, BRE Digest 329, 1988.

Replacing wall ties, Building Research Establishment, BRE Digest 401, 1995.

Connecting walls and floors, Building Research Establishment, BRE Good Building Guide 29: Parts 1 and 2, 1997.

The design of windposts and parapet posts, Ancon Stainless Steel Fixings Ltd., Design Briefing No. 4, Sheffield.

Treating dampness in basements, Building Research Establishment, BRE Good Repair Guide 23, 1999.

Removing internal walls in older buildings, Building Research Establishment, BRE Good Building Guide 20, 1998.

Improving energy efficiency: thermal insulation, Building Research Establishment, BRE Good Repair Guide 26, 1999.

Insulated external cladding systems, Building Research Establishment, BRE Good Building Guide 31, 1999.

Improved standards of insulation in cavity walls with an outer leaf in facing brickwork, Ford and Durose, Brick Development Association, BDA Publ. DN11, 1982.

Improving the sound insulation of separating walls and floors, Building Research Establishment, BRE Digest 293, 1985.

Index